CW01674469

Isetta

Gold Portfolio
1953-1964

Compiled by R M Clarke

ISBN 9781855203815

BROOKLANDS BOOKS LTD.
P.O. BOX 146, COBHAM,
SURREY, KT11 1LG. UK
sales@brooklands-books.com

ISGP

www.brooklands-books.com

Printed in China

Brooklands Books

ROAD TEST SERIES

Abarth Gold Portfolio 1950-1971
Alfa Romeo Giulietta Gold Portfolio 1954-1965
Alfa Romeo Giulia Berlina Lim. Edit. Extra 1962-76
Alfa Romeo Giulia Coupes Lim. Edit. Ultra 1973-76
Alfa Romeo Alfasud 1972-1984
Alfa Romeo Alfetta Gold Portfolio 1972-1987
Alfa Romeo Spider Ultimate Portfolio 1966-1994
Alfa Romeo Spider & GTV Perf. Port. 1995-2005
Allard Limited Edition Ultra
Alpine Renault Ultimate Portfolio 1958-1995
Alvis Gold Portfolio 1919-1967
AMC Rambler Limited Edition Extra 1956-1969
AMX & Javelin Gold Portfolio 1968-1974
Armstrong Siddeley Gold Portfolio 1945-1960
Aston Martin Gold Portfolio 1921-1947
Aston Martin Ultimate Portfolio 1948-1968
Aston Martin Ultimate Portfolio 1968-1980
Aston Martin Ultimate Portfolio 1981-1993
Aston Martin Ultimate Portfolio 1994-2006
Audi Quattro Gold Portfolio 1980-1991
Audi Quattro Takes On The Competition
Audi TT Performance Portfolio 1998-2006
Austin-Healey 100 & 100/6 Gold Port. 1952-1959
Austin-Healey 3000 Ultimate Portfolio 1959-1967
Austin-Healey Sprite Gold Portfolio 1958-1971
Bentley & Rolls-Royce Portfolio 1990-2002
Berkeley Sportscars Limited Edition
BMW 6 & 8 Cyl. Cars Limited Edition 1935-1960
BMW 700 Limited Edition 1959-1965
BMW 1600 Collection No. 1 1966-1981
BMW 2002 Ultimate Portfolio 1968-1976
BMW 6 Cylinder Coupes & Saloons Gold P. 1969-1976
BMW 316, 318, 320 (4 cyl.) Gold Port. 1975-1990
BMW 320, 323, 325 (6 cyl.) Gold Port. 1977-1990
BMW 3 Series Gold Portfolio 1991-1997
BMW M3 Ultimate Portfolio 1986-2006
BMW M5 Gold Portfolio 1980-2003
BMW 5 Series Gold Portfolio 1988-1995
BMW 6 Series Ultimate Portfolio 1976-1989
BMW 7 Series Performance Portfolio 1977-1986
BMW 7 Series Performance Portfolio 1986-1993
BMW 8 Series Performance Portfolio
BMW X5 Limited Edition Extra 1999-2006
BMW Alpina Performance Portfolio 1967-1987
BMW Alpina Performance Portfolio 1988-1998
BMW Z3, M Coupe & M Roadster Gold Port. 1996-02
Borgward Isabella Limited Edition
Bristol Cars Portfolio
Buick Performance Portfolio 1947-1962
Buick Muscle Portfolio 1963-1973
Buick Riviera Performance Portfolio 1963-1978
Cadillac Performance Portfolio 1948-1958
Cadillac Performance Portfolio 1959-1966
Cadillac Eldorado Performance Portfolio 1967-1978
Cadillac Allante Limited Edition Extra
Impala & SS Muscle Portfolio 1958-1972
Corvair Performance Portfolio 1959-1969
El Camino & SS Muscle Portfolio 1959-1987
Chevy II & Nova SS Gold Portfolio 1962-1974
Chevelle & SS Gold Portfolio 1964-1972
Camaro Muscle Portfolio 1967-1973
Blazer & Jimmy Limited Edition Extra 1969-1982
Blazer & Jimmy Limited Edition Extra 1983-1994
Camaro Performance Portfolio 1993-2000
Chevrolet Corvette Gold Portfolio 1953-1962
Chevrolet Corvette Sting Ray Gold Port. 1963-1967
Chevrolet Corvette Gold Portfolio 1968-1977
Chrysler Imperial Gold Portfolio 1951-1975
Valiant 1960-1962
PT Cruiser Performance Portfolio
Citroen Traction Avant Limited Edition Premier
Citroen 2CV Ultimate Portfolio 1948-1990
Citroen DS & ID 1955-1975
Citroen DS & ID Gold Portfolio 1955-1975
Citroen SM Limited Edition Extra 1970-1975
Shelby Cobra Gold Portfolio 1962-1969
Crosley & Crosley Specials Limited Edition
Cunningham Automobiles 1951-1955
Datsun Roadsters Performance Portfolio 1960-71
Datsun 240Z & 260Z Gold Portfolio 1970-1978
DeLorean Gold Portfolio 1977-1995
De Soto Limited Edition 1952-1960
Dodge Limited Edition 1949-1959
Dodge Dart Limited Edition Extra 1960-1976
Dodge Muscle Portfolio 1964-1971
Charger Muscle Portfolio 1966-1974
ERA Gold Portfolio 1934-1994
Facel Vega Limited Edition Extra 1954-1964
Ferrari Limited Edition 1947-1957
Ferrari Dino Limited Edition Extra 1965-1974
Ferrari 308 & Mondial Ultimate Portfolio 1975-85
Ferrari 328 348 Mondial Ultimate Portfolio 1986-94
Ferrari F355 & 360 Gold Portfolio 1995-2004
Fiat 500 1936-1972 Road Test Portfolio
Fiat 600 & 850 Gold Portfolio 1955-1972
Fiat Dino Road Test Portfolio
Fiat 124 Spider Performance Portfolio 1966-1985
Fiat X1/9 Gold Portfolio 1973-1989
Ford Consul, Zephyr, Zodiac Mk. I & II 1950-1962
Ford Zephyr, Zodiac, Executive Mk. III & IV 1962-1971
High Performance Capris Gold Portfolio 1969-1987
Capri Muscle Portfolio 1974-1987
High Performance Fiestas 1979-1991
Ford Escort RS & Mexico Limited Edition 1970-1979
High Performance Escorts Mk. II 1975-1980
High Performance Escorts 1980-1985
High Performance Escorts 1985-1990
Ford Thunderbird Performance Portfolio 1955-1957
Ford Thunderbird Performance Portfolio 1958-1963
Ford Thunderbird Performance Portfolio 1964-1976
Ford Fairlane Performance Portfolio 1955-1970
Ford Ranchero Muscle Portfolio 1957-1979
Edsel Limited Edition 1957-1960
Ford Galaxie & LTD Gold Portfolio 1960-1976
Falcon Performance Portfolio 1960-1970
Ford GT40 & GT Ultimate Portfolio 1964-2006
Ford Torino Performance Portfolio 1968-1974
Ford Bronco 4x4 Performance Portfolio 1966-1977
Ford Bronco 1978-1988
Shelby Mustang Ultimate Portfolio 1965-1970
Mustang Muscle Portfolio 1967-1973
High Performance Mustang IIs 1974-1978
Mustang 5.0L Muscle Portfolio 1982-1993
Mustang 5.0L Takes On The Competition
Ginetta Cars Limited Edition Ultra 1958-2007
Goggomobil Limited Edition
Honda S500 • S600 • S800 Limited Edition 1962-1970
Honda CRX 1983-1987
Honda S2000 Performance Portfolio 1999-2008
Hudson Performance Portfolio 1946-1957
International Scout Gold Portfolio 1961-1980
Isetta Gold Portfolio 1953-1964
ISO & Bizzarrini Limited Edition Ultra 1962-1974
Jaguar and SS Gold Portfolio 1931-1951
Jaguar XK120 • 140 • 150 Gold Portfolio 1948-1960
Jaguar C-Type & D-Type Gold Portfolio 1951-1960
Jaguar Mk 1 & Mk 2 1955-1969 Road Test Portfolio
Jaguar E-Type Ultimate Portfolio 1961-1975
Jaguar XJ6 Series I & II Gold Portfolio 1968-1979
Jaguar XJ6 Series III Perf. Portfolio 1979-1986
Jaguar XJS Gold Portfolio 1975-1988
Jaguar XJ-S V12 Ultimate Portfolio 1988-1996
Jaguar XK8 & XKR Performance Portfolio 1996-2005
Jeep CJ-5 Limited Edition 1960-1975
Jeep CJ-5 & CJ-7 4x4 Perf. Portfolio 1976-1986
Jeep Wagoneer Performance Portfolio 1963-1991
Jeep J-Series Pickups 1970-1982
Jeepster & Commando Limited Edition 1967-1973
Jeep Cherokee & Comanche Pickups P. P. 1984-91
Jeep Wrangler 4x4 Performance Portfolio 1987-99
Jeep Cherokee & Grand Cherokee 4x4 P. P. 1992-98
Jensen Interceptor Ultimate Portfolio 1966-1992
Jensen - Healey Limited Edition 1972-1976
Kaiser - Frazer Limited Edition 1946-1955
Lagonda Gold Portfolio 1919-1964
Lancia Aurelia & Flaminia Gold Portfolio 1950-1970
Lancia Fulvia Gold Portfolio 1963-1976
Lancia Beta Gold Portfolio 1972-1984
Lancia Stratos Limited Edition Extra
Lancia Delta & integrale Ultimate Portfolio
Land Rover Series I, II & IIA Gold Portfolio 1948-71
Land Rover Series III 4x4 Perf. Portfolio 1971-1985
Land Rover 90 110 Defender Gold Portfolio 1983-94
Land Rover Discovery Perf. Port. 1989-2000
Lamborghini Performance Portfolio 1964-1976
Lamborghini Performance Portfolio 1977-1989
Lamborghini Gold Portfolio 1990-2004
Lincoln Gold Portfolio 1949-1960
Lincoln Continental Performance Portfolio 1961-1969
Lincoln Continental 1969-1976
Lotus Sports Racers Portfolio - covering 1951-1965
Lotus Seven Gold Portfolio 1957-1973
Lotus Elite Limited Edition 1957-1964
Lotus Elan Ultimate Portfolio 1962-1974
Lotus Elan & SE 1989-1992
Lotus Europa Gold Portfolio 1966-1975
Lotus Elite & Eclat 1974-1982
Lotus Elise & Exige Gold Portfolio 1995-2005
Marcos Coupés & Spyders Gold Portfolio 1960-1997
Maserati Cars Performance Portfolio 1957-1970
Maserati Cars Performance Portfolio 1971-1982
Maserati Cars Performance Portfolio 1982-1998
Maserati Cars Ultimate Portfolio 1999-2007
Matra Limited Edition 1965-1983
Mazda Miata MX-5 Performance Portfolio 1989-1997
Mazda Miata MX-5 Performance Portfolio 1998-2005
Mazda Miata MX-5 Takes On The Competition
McLaren F1 • GTR • LM Sportscar Perf. Portfolio
Mercedes 190 & 300 SL 1954-1963
Mercedes S & 600 Limited Edition Extra 1965-1972
Mercedes S Class 1972-1979
Mercedes S Class Limited Edition Extra 1980-1991
Mercedes 230 • 250 • 280SL Gold Portfolio 1963-1971
Mercedes-Benz SLs & SLCs Ultimate Port. 1971-89
Mercedes SLs Performance Portfolio 1989-1994
Mercedes G-Wagen Gold Portfolio 1981-2005
Mercedes 190 Limited Edition Extra 1983-1993
Mercedes CLK & SLK Limited Edition 1996-2000
Mercedes AMG Gold Portfolio 1983-1999
Mercedes AMG Ultimate Portfolio 2000-2006
Mercury Gold Portfolio 1947-1966
Mercury Comet & Cyclone Lim. Edit. Extra 1960-75
Cougar Muscle Portfolio 1967-1973
Messerschmitt Gold Portfolio 1954-1964
MG Gold Portfolio 1929-1939
MG TA & TC Gold Portfolio 1936-1949
MG TD & TF Gold Portfolio 1949-1955
MGA & Twin Cam Gold Portfolio 1955-1962
MG Midget Gold Portfolio 1961-1979
MGB Roadsters 1962-1980
MGB MGC & V8 Gold Portfolio 1962-1980
MGC & MGB GT V8 Limited Edition
MGF & TF Performance Portfolio 1995-2005
Mini Gold Portfolio 1959-1969
Mini Gold Portfolio 1969-1980
Mini Gold Portfolio 1981-1997
High Performance Minis Gold Portfolio 1960-1973
Mini Cooper Gold Portfolio 1961-1971
Mini Moke Ultimate Portfolio 1964-1994
Mini Performance Portfolio 2001-2006
Starion & Conquest Performance Portfolio 1982-1990
Mitsubishi 3000GT & Dodge Stealth P.P. 1990-1999
Morgan Three-Wheelers Ultimate Portfolio 1909-1952
Morgan Four-Wheelers Ultimate Portfolio 1936-1967
Morgan Ultimate Portfolio 1968-1990
Morgan Ultimate Portfolio 1991-2009
Nash Limited Edition Extra 1949-1957
Nash-Austin Metropolitan Gold Portfolio 1954-1962
NSU Ro80 Limited Edition
NSX Performance Portfolio 1989-1999
Oldsmobile 1948-1963 Limited Edition Premier
Oldsmobile Muscle Portfolio 1964-1971
Cutlass & 4-4-2 Muscle Portfolio 1964-1974
Opel GT Ultimate Portfolio 1968-1973
Opel Manta Limited Edition 1970-1975
Pantera Ultimate Portfolio 1970-1995
Panther Gold Portfolio 1972-1990
Plymouth Limited Edition 1950-1960
Plymouth Fury Limited Edition Extra 1956-1976
Barracuda Muscle Portfolio 1964-1974
Plymouth Muscle Portfolio 1964-1971
Pontiac 1946- 1963 Limited Edition Premier
High Performance Firebirds 1982-1988
Firebird & Trans Am Performance Portfolio 1993-00
Pontiac Fiero Performance Portfolio 1984-1988
Porsche Sports Racing Cars UP 1952-1968
Porsche 917 • 935 • 956 • 962 Gold Portfolio
Porsche 912 Limited Edition Extra
Porsche 365 Ultimate Portfolio 1952-1965
Porsche 911 1965-1969
Porsche 911 1973-1977
Porsche 911 SC & Turbo Gold Portfolio 1978-1983
Porsche 911 Carrera & Turbo Gold Port. 1984-1989
Porsche 911 Ultimate Portfolio 1990-1997
Porsche 911 Takes On The Competition 1990-1997
Porsche 911 Ultimate Portfolio 1998-2004
Porsche 914 Ultimate Portfolio
Porsche 924 Gold Portfolio 1975-1988
Porsche 928 Gold Portfolio 1977-1995
Porsche 928 Takes On The Competition
Porsche 944 Ultimate Portfolio
Porsche 968 Limited Edition Extra
Porsche Boxster Ultimate Portfolio 1996-2004
Railton & Brough Superior Gold Portfolio 1933-1950
Range Rover Gold Portfolio 1970-1985
Range Rover Gold Portfolio 1985-1995
Range Rover Performance Portfolio 1995-2001
Range Rover Takes on the Competition
Riley Gold Portfolio 1924-1939
Rolls-Royce Silver Clould & Bentley S Ultimate Port.
Rolls-Royce Silver Shadow Ultimate Port. 1965-80
Rover P4 1949-1959
Rover 2000 & 2200 1963-1977
Studebaker Ultimate Portfolio 1946-1966
Studebaker Hawks & Larks Lim. Edit. Premier 1956-66
Avanti Limited Edition Extra 1962-1991
Subaru Impreza Turbo Limited Edition Extra 1994-00
Subaru Impreza WRX Performance Port. 2001-2005
Sunbeam Alpine Limited Edition Extra 1959-1968
Sunbeam Tiger Limited Edition Extra 1964-1967
Suzuki SJ Gold Portfolio 1971-1997
Vitara, Sidekick & Geo Tracker Perf. Port. 1988-1997
Toyota Land Cruiser Gold Portfolio 1956-1987
Toyota Land Cruiser 1988-1997
Toyota Supra Performance Portfolio 1982-1998
Toyota MR2 Ultimate Portfolio 1984-2006
Toyota MR2 Takes On The Competition
Triumph TR2 & TR3 Gold Portfolio 1952-1961
Triumph TR4, TR5, TR250 1961-1968
Triumph TR6 Gold Portfolio 1969-1976
Triumph Herald 1959-1971
Triumph Vitesse 1962-1971
Triumph Spitfire Gold Portfolio 1962-1980
Triumph 2000, 2.5, 2500 1963-1977
Triumph GT6 Gold Portfolio 1966-1974
Triumph Stag Gold Portfolio 1970-1977
TVR Limited Edition Ultra 1958-1985
TVR Performance Portfolio 1986-1994
TVR Performance Portfolio 1995-2000
TVR Performance Portfolio 2000-2005
VW Beetle Gold Portfolio 1935-1967
VW Beetle Gold Portfolio 1968-1991
VW Bus Camper Van Perf. Portfolio 1954-1967
VW Bus Camper Van Perf. Portfolio 1968-1979
VW Bus Camper Van Perf. Portfolio 1979-1991
VW Karmann Ghia Gold Portfolio 1955-1974
VW Scirocco 1974-1981
VW Golf GTI Limited Edition Extra 1976-1991
VW Corrado Limited Edition Premier 1989-1995
Volvo PV444 & PV544 Perf. Portfolio 1945-1965
Volvo 120 Amazon Ultimate Portfolio
Volvo 1800 Ultimate Portfolio 1960-1973
Volvo 140 & 160 Series Gold Portfolio 1966-1975
Forty Years of Selling Volvo
Westfield Performance Portfolio 1982-2004

RACING & THE LAND SPEED RECORD

The Land Speed Record 1898-1919
The Land Speed Record 1920-1929
The Land Speed Record 1930-1939
The Land Speed Record 1940-1962
The Land Speed Record 1963-1999
Can-Am Racing 1966-1969
Can-Am Racing 1970-1974
Can-Am Racing Cars 1966-1974
The Carrera Panamericana Mexico - 1950-1954
Le Mans - The Bentley & Alfa Years - 1923-1939
Le Mans - The Jaguar Years - 1949-1957
Le Mans - The Ferrari Years - 1958-1965
Le Mans - The Ford & Matra Years - 1966-1974
Le Mans - The Porsche Years - 1975-1982
Le Mans - The Porsche & Jaguar Years - 1983-91
Le Mans - The Porsche & Peugeot Years - 1992-99
Mille Miglia - The Alfa & Ferrari Years - 1927-1951
Mille Miglia - The Ferrari & Mercedes Years - 1952-57
Targa Florio - The Porsche & Ferrari Years - 1955-1964
Targa Florio - The Porsche Years - 1965-1973

RESTORATION & GUIDE SERIES

BMW 2002 - A Comprehensive Guide
BMW 02 Restoration Guide
BMW E30 - 3 Series Restoration Bible
Classic Camaro Restoration
Engine Swapping Tips & Techniques
Ferrari Life Buyer's Portfolio
Land Rover Restoration Portfolio
PC on Land Rover Series I Restoration
Lotus Elan Restoration Guide
MG T Series Restoration Guide
MGA Restoration Guide
PC on Midget/Sprite Restoration
PC on MGB Restoration
Mustang Restoration Tips & Techniques
Practical Gas Flow
Restoring Sprites & Midgets an Enthusiast's Guide
SU Carburetters Tuning Tips & Techniques
PC on Sunbeam Rapier Restoration
The Great Classic Muscle Cars Compared
Weber Carburettors Tuning Tips and Techniques

MILITARY VEHICLES

Complete WW2 Military Jeep Manual
Dodge WW2 Military Portfolio 1940-1945
German Military Equipment WW2
Hail To The Jeep
Combat Land Rover Portfolio No. 1
Land Rover Military Portfolio
Military & Civilian Amphibians 1940-1990
Off Road Jeeps Civilian & Military 1944-1971
US Military Vehicles 1941-1945
Standard Military Motor Vehicles-TM9-2800 (WW2)
VW Kubelwagen Military Portfolio 1940-1990
WW2 Allied Vehicles Military Portfolio 1939-1945
WW2 Jeep Military Portfolio 1941-1945

ROAD & TRACK SERIES

Road & Track on Aston Martin 1962-1990
Road & Track on Austin Healey 1953-1970
Road & Track on BMW Cars 1966-1974
Road & Track BMW M Series Portfolio 1979-2002
R & T BMW Z3, M Coupe & M Roadster Port. 96-02
R & T Camaro & Firebird Portfolio 1993-2002
Road & Track on Corvette 1968-1982
Road & Track on Corvette 1982-1986
Road & Track on Corvette 1986-1990
Road & Track Corvette Portfolio 1997-2002
Road & Track Dodge Viper Portfolio 1992-2002
Road & Track on Ferrari 1975-1981
Road & Track on Ferrari 1984-1988
Road & Track Ferrari V-12 Portfolio 1992-2002
Road & Track Ferrari F355 360 F430 Portfolio 95-06
Road & Track on Fiat Sports Cars 1968-1987
Road & Track on Jaguar 1950-1960
Road & Track on Jaguar 1961-1968
Road & Track on Jaguar 1968-1974
Road & Track on Jaguar 1974-1982
Road & Track on Jaguar 1983-1989
R & T Jaguar XJ-S - XK8 - XKR Portfolio 1975-2003
Road & Track MX-5 Miata Portfolio 1989-2002
Road & Track on Mercedes 1952-1962
Road & Track on Mercedes 1963-1970
Road & Track on Mercedes 1971-1979
R & T Mercedes SL - SLK - CLK Portfolio 1990-2003
Road & Track on MG Sports Cars 1949-1961
Road & Track on MG Sports Cars 1962-1980
Road & Track Mustang Portfolio 1994-2002
Road & Track Nissan 300ZX & 350Z Portfolio 1984-03
Road & Track on Porsche 1951-1967
Road & Track on Porsche 1972-1975
Road & Track on Porsche 1975-1978
Road & Track on Porsche 1979-1982
Road & Track on Porsche 1985-1988
Road & Track Porsche 928 Portfolio 1977-1994
Road & Track Porsche 911 Portfolio 1990-1997
R & T on Rolls Royce & Bentley 1950-1965
R & T on Rolls Royce & Bentley 1966-1984
R & T on Toyota Sports & GT Cars 1966-1984
R & T on Triumph Sports Cars 1967-1974
R & T on Triumph Sports Cars 1974-1982
Road & Track on Volkswagen 1951-1968
Road & Track on Volkswagen 1968-1978
Road & Track on Volvo 1957-1974
Road & Track on Volvo 1977-1994
Road & Track - Best of PS
Road & Track - Peter Egan Side Glances 1983-1992
Road & Track - Peter Egan Side Glances 1992-1997
Road & Track - Peter Egan Side Glances 1998-2002
Road & Track - Peter Egan Side Glances 2002-2006

CAR AND DRIVER SERIES

Car and Driver on BMW 1957-1977
Car and Driver on Corvette 1978-1982
Car and Driver on Corvette 1983-1988
Car and Driver on Ferrari 1955-1962
Car and Driver on Ferrari 1963-1975
Car and Driver on Porsche 1955-1962
Car and Driver on Porsche 1963-1970
Car and Driver on Porsche 1970-1976
Car and Driver on Porsche 1977-1981
Car and Driver on Porsche 1982-1986

HOT ROD 'ENGINE' SERIES

Chevy 265 & 283
Chevy 302 & 327
Chevy 348 & 409
Chevy 396 & 427
Chevy 454 thru 512
Chevy Monster Big Blocks
Chrysler Hemi
Chrysler 273, 318, 340 & 360
Chrysler 361, 383, 400, 413, 426 & 440
Ford 289, 302, Boss 302 & 351W
Ford 351C & Boss 351
Ford Small Block
Ford Big Block

MOTORCYCLES

To see our range of over 70 titles visit www.brooklands-books.com

.75/05Z92000

Brooklands Books

CONTENTS

Brooklands Books

ACKNOWLEDGEMENTS

About three years ago we published one of our regular 100 page books on the Isetta. It sold very well and because we have had a great deal of support from owners and followers of the marque we have decided to upgrade it and reissue it as a Gold Portfolio containing a further 72 pages of new text. The original Isetta book was in some ways an unusual book for Brooklands to publish, not least because it really covers more than one make of car: the Isetta was built by different companies in no fewer than four different countries. Nevertheless, the format we have adopted is the one familiar to buyers of the Brooklands series. To magazine articles published about the cars when they were new, we have added some more recent appraisals and several contemporary publicity items.

The magazine articles would not be here at all of we were not fortunate enough to enjoy the understanding and co-operation of those who hold the copyright to them. Our thanks therefore go to the publishers of *Auto Age, Auto Sport Review, Autocar, Autosport, BMW, Bubble Notes, Car, Car and Driver, Car Collector, Car Life, Car South Africa, Cars of the World, Classic and Sportscar, Coast Car Collector, Foreign Car Shoppers Guide, Isetta G.B., ISO, Modern Motor, Motor, Motor Cycle, Motor Guide, Motor Life, Motor Sport, Motorsport, Motor Trend, Old Motor, Popular Mechanics, Road & Track, Small Car Guide, Small Cars Illustrated, Small Car Parade, Special Interest Autos, U.S. Isetta Importers, Velam, Wheels, World Car Guide* and *World Wide Automobiles*. Our thanks also go to David Kayser and Marilyn Felling who generously loaned us stories and advertising material on the Isetta. We also received help from three sources with regard to our cover pictures. For our front cover photograph our thanks go to Peter Egan and Road & Track magazine and the story behind this picture and car can be found on page 164. For the three right hand photographs on our back cover our thanks go to Dave Watson the librarian of the Isetta Owners Club for the use of these. Finally our thanks go to John Wetzel (Isetta John) who supplied us with the top left hand picture on our back cover which shows Ernie Freestones beautifully restored car which I see each year on display at Hershey.

R M Clarke

Isetta was the bright idea of Renzo Rivolta, a wealthy Italian motorcycle and motor scooter manufacturer. Introduced in 1953, it was initially manufactured by his company, Iso SpA of Milan, and had a 250cc two-stroke twin-cylinder engine. But Rivolta had bigger ambitions, and within a couple of years he had sold manufacturing licenses to BMW in Germany and Velam in France. All that was left for him to do then was to sit back and reap the profits. In fact, he stopped making the vehicle himself, and turned instead to exotic sports cars.

BMW hoped that sales of this tiny car would underpin slow sales of their luxury saloons and sports models. Their version appeared in April 1955, and was powered by their own 247cc motorcycle engine. Within a year, this original Isetta 250 model was joined by an Isetta 300, with a more powerful 297cc engine.

Meanwhile, the Isetta had also entered production in France, and in 1957 it also appeared in Britain. Isetta of Great Britain, Ltd. shrewdly cashed in on the trend towards more fuel efficient cars which resulted from the Suez Canal crisis. Holding their manufacturing license from BMW the company built a version of the car which had a single rear wheel instead of the narrow-track twin rear wheels of its parent.

BMW now broadened its horizons with the 600 model, a wheel-at-each-corner, four-seat, long wheelbase development of the original Isetta. This appeared in December 1957, but never attained the popularity of the original model and was withdrawn after just two years. In its place, BMW introduced the 700, which owed nothing to the Isetta tradition.

Although, the Isetta was remarkably popular in the later 1950's, the fashion for cars of its type died out after BMC introduced the enormously successful Mini in 1959. BMW stopped manufacturing this range in 1962, and production in Britain ceased just two years later. Within a short period, the public had totally lost interest in the Isetta, and it was not until 20 years or more later that enthusiasts began to recognise the interesting nature of these little cars. Today, some 30 years after the last one was built, this is a welcome reminder of the Isetta's colourful, eccentric, but eminently practical approach to motoring.

James Taylor

EGG WITH WHEELS

Rome:

THOUGH at first glance it might appear the Italian auto industry had laid an egg with this one—almost literally—you have to give them credit for imagination, at least.

The Isetta, which is scheduled to go on sale in Italy next month, is a highly-ingenious solution to the problem of the absolute minimum car. Designed and built by the Iso company of Milan, manufacturers of the popular Iso-Moto motor scooter, it is a four-wheeled vehicle powered by a two-stroke engine of only 198 cc (which has a double cylinder and single combustion chamber). This odd item churns out a mighty 9.5 horsepower at 4750 rpm on a compression ratio of 6.5 to 1 and supposedly goes 55 mph.

The engine is located in the rear, on the right side to balance the driver's weight. It is air-cooled by a forced stream of air. The rear wheels are very close together and there are no normal half-shafts for the drive. Instead the four-speed transmission transfers the motion to a single rear axle shaft through a double chain drive. Rear suspension is by leaf spring and the independent front wheels are suspended by special rubber cushions. Baggage space, spare tire and battery all are in the rear.

The Isetta carries up to three passengers, claims 71 miles per gallon and weighs only 682 pounds. It has an optional heating system and will sell for $650. Once a satisfactory rate of production is reached, the cars will be exported just as the firm's scooters have been—and particular plans are being made for U.S. sales.

Production plans call for 50 cars a day from a specially-built factory and as production goes up the price will be reduced. However, this policy has met opposition from the dominant Fiat company and it is likely that Iso will not be allowed to cut the price too much.

The most unusual feature of the Isetta is the single door, which forms the front of the car and opens forward. The steering wheel and its column are fixed to the inside of the front door and open with it by means of a universal joint. Despite the inclination of the hinges, the door automatically stays open in any position.

This little car has all the equipment of a normal-sized car—hydraulic brakes, headlights, bumpers, windshield wiper, starter, clutch, accelerator and ventilating windows. The gear lever and handbrake lever are located on the left side of the driver.

The body is monobuilt of pressed steel, without a frame, and steel tubing supports the windows and roof. There is a small canvas roof that can be snapped off.

It is claimed the Isetta is good for at least 60,000 miles of normal driving without major repairs.

●●●

By Melton Davis

The Motor Road Test No. 28/54 (Continental)

Make: Iso **Type:** Isetta

Makers: Iso S.p.A., 16, Corso Porta Nuova, Milan, Italy

(Test car loaned by Anc. Ets. Pilette S.A., 510, Chaussée de Louvain, Brussels)

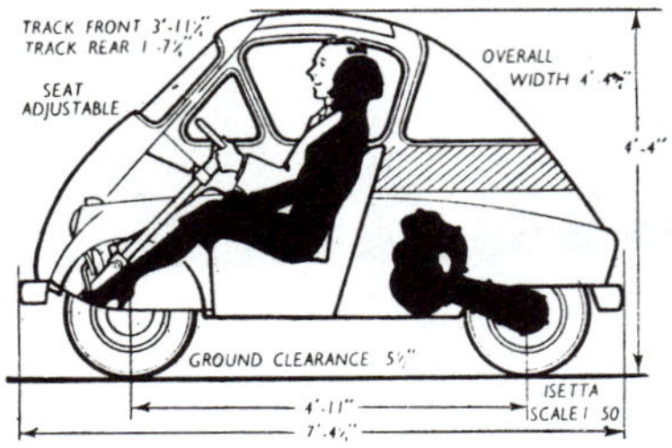

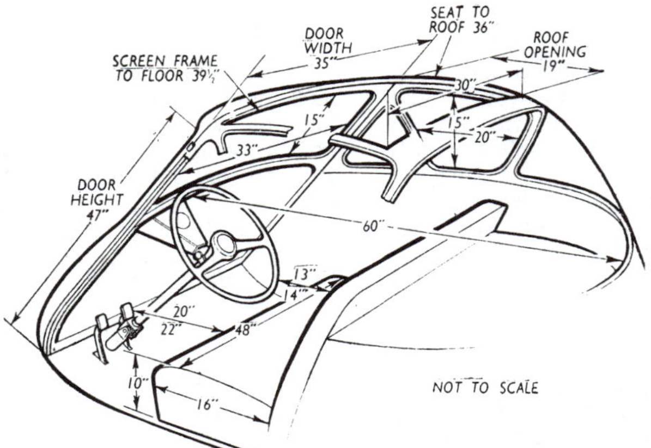

Test Data

CONDITIONS: *Hot, dry weather with moderate breeze. Smooth concrete surface. Standard and premium-grade Belgian pump fuels. Tests made with 2 people and box of test gear in car.*

INSTRUMENTS:

Speedometer at 30 m.p.h.	11% fast
Distance recorder	5% fast

MAXIMUM SPEEDS

Flying Quarter Mile

Mean of six opposite runs ..	44.9 m.p.h.
Best time equals	46.2 m.p.h.

Speed in Gears

Max. speed in 3rd gear	34 m.p.h.
Max. speed in 2nd gear	22 m.p.h.
Max. speed in 1st gear	12 m.p.h.

FUEL CONSUMPTION

77.5 m.p.g. at constant 20 m.p.h.
72.5 m.p.g. at constant 30 m.p.h.
57.5 m.p.g. at constant 40 m.p.h.

Overall consumption for 219 miles, driven hard, 4.35 gallons = 50.4 m.p.g.

Fuel tank capacity 2.9 gallons (including 0.6-gallon reserve).

ACCELERATION TIMES Through Gears

0-20 m.p.h...	6.0 sec.
0-30 m.p.h...	13.8 sec.
0-40 m.p.h...	35.8 sec.
Standing Quarter Mile	32.5 sec.

ACCELERATION TIMES on Two Upper Ratios

	Top	3rd
10-30 m.p.h.	—	13.7 sec.
20-40 m.p.h.	38.6 sec.	—

WEIGHT

Unladen kerb weight	6¾ cwt.
Front/rear weight distribution ..	56/44
Weight laden as tested	10¼ cwt.

HILL CLIMBING (at steady speeds)

Max. gradient on top gear	1 in 35	(Tapley 65 lb./ton)
Max. gradient on 3rd gear	1 in 13	(Tapley 170 lb./ton)
Max. gradient on 2nd gear	1 in 7.2	(Tapley 305 lb./ton)

BRAKES at 30 m.p.h.

0.71 g retardation (=42½ ft. stopping distance) with 50 lb. pedal pressure
0.15 g retardation (=200 ft. stopping distance) with 25 lb. pedal pressure

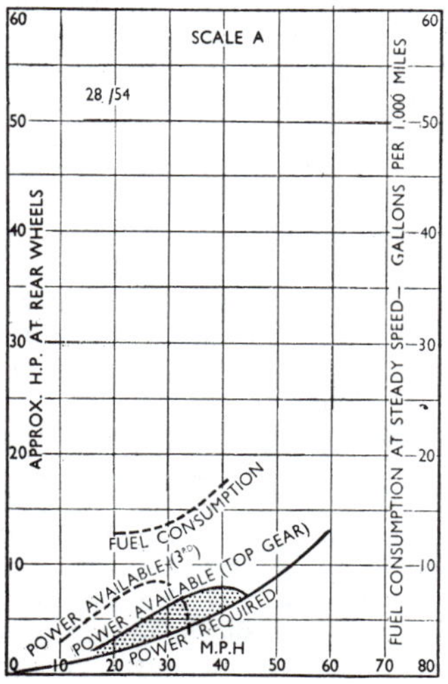

Drag at 10 m.p.h. .. .. 29 lb.
Drag at 60 m.p.h., by extrapolation, approx. .. .. 83 lb.
Specific Fuel Consumption when cruising at 80% of maximum speed (i.e. 35.9 m.p.h.) on level road, based on power delivered to rear wheels .. .. 0.93 pints/b.h.p./hr.

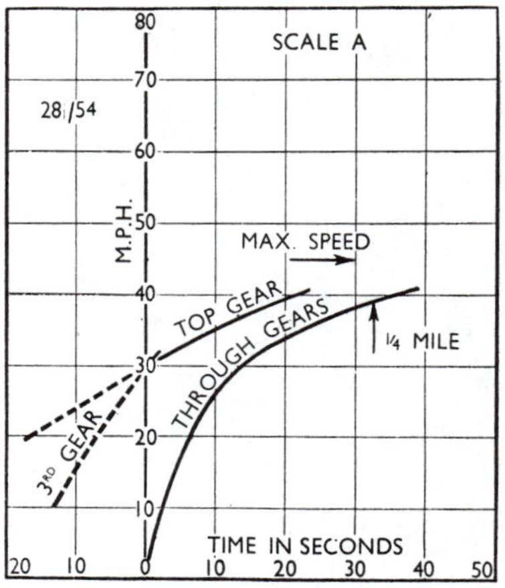

Maintenance

Engine oil tank: 5½ pints, S.A.E. 30. **Gearbox:** 1 pint, S.A.E. 40. **Rear Axle:** 1 pint, S.A.E. 40. **Steering gear:** Grease. **Radiator:** Nil (air-cooled). **Chassis lubrication:** By grease gun every 600 miles to 6 points. **Ignition timing:** ⅜ in. before B.T.D.C. (rear piston), fully advanced. **Spark plug gap:** 0.023 in.-0.028 in. **Contact breaker gap:** 0.016 in.-0.024 in. **Valve timing:** Not applicable. **Tappet clearances:** Not applicable. **Front wheel toe-in:** 0.16-0.20 mm. at rims. **Camber angle:** 1½°. **Tyre pressures:** 10-12 lb./sq. in. **Brakes:** Hydraulic brake fluid. **Battery:** 12 volt, 24 amp./hr. **Lamp bulbs:** Head lamps 30/30 watt; Side and tail lamps 3/20 watt; Number plate lamp 3 watt; Instrument light 1.5 watt.

Ref. 1/3/54

—The ISO ISETTA

An Unorthodox Italian 4-wheeled Runabout of Exceptionally Compact Form

COMPACT dimensions are obtained by use of a forward seating position reached through a front door. A folding canvas roof panel is provided, and there is a spacious luggage shelf above the rear-mounted engine and narrow-track rear axle.

DESIGNED and manufactured in Italy, but already being assembled in Belgium and possibly to be produced in Britain also next year, the Isetta is a boldly unconventional vehicle which fulfils two main purposes. It provides transportation of a kind which is cheaper than conventional motoring but more comfortable by far than motorcycling. Furthermore, it carries two people in fair style and comfort in a four-wheeled vehicle which occupies only half as much area as a popular 8 h.p. car when running or parked in crowded city streets.

It is likely that the majority of readers, even those familiar with most of the world's cars, will have turned their attention to the accompanying pictures before reading this report. Apart from mention in reports of the Turin motor shows (and, incidentally, in the report of the Mille Miglia race, where a team of these little machines were victorious under the Index of Performance) the Isetta has not hitherto been greatly described. A brief résumé of its design may therefore be worth including.

For such a small car the frame is extremely rigid, the main members of box section with tubular cross-bracing, and extra stiffening is provided by the light tubular scaffolding which carries the body panels. Rubber mountings support the rear-mounted, double-cylinder, two-stroke engine, which some may recall as following the same principles as the old Trojan two-stroke, with a common combustion chamber for the two parallel cylinders, and connecting rods working on the same throw of the crankshaft. The inlet port is in one cylinder, the exhaust in the other, an arrangement which gives particularly good scavenging. The drive, through a four-speed synchromesh gearbox, is taken by duplex chain to a short rear axle without differential gears, the axle and chain case swinging as one unit, with quarter-elliptic springs. At the front, Dubonnet suspension is used with a springing medium, not yet made final, of either coil springs or rubber in compression.

Compactness has been valued far more highly than cheapness in the conception of this unique vehicle. Elaboration of detail has not been shirked when it could provide improved results, but convention has been flouted completely in order to minimize overall length without sacrificing comfort. The basic difference between the Isetta and any other vehicle is that entry is from the front in the manner of the hansom cab, instead of being from the side as is nowadays usual. To make this layout convenient, there is the further unexpected feature of a steering wheel and column which swing forwards with the counterbalanced door, to give really easy access to the driving seat. Two, or at a pinch three, people occupy a single bench seat 48 inches wide, their feet extending actually ahead of the front wheels, and the power unit is packed into a very small space behind them. An overall width of 4 ft. 4¾ in. is commendably compact, but an overall length of 7 ft. 4½ in. is a full 4 feet less than the corresponding dimension of the smallest British car.

A Serious Design

Can a car of such small size be comfortable and useful? After covering some hundreds of miles in a far-from-new Isetta, our answer is a clear "yes." Any vehicle has its limitations, but this one offers enough comfort and enough performance for very many requirements, is inexpensive to run, and could make a substantial contribution to the solution of congestion and parking problems.

No high speeds were expected during this test, but the example driven proved capable of a genuine 45 m.p.h., and covered more than 50 miles on each gallon of

In Brief

Price in Belgium 41,500 B.fr., equals approx. £300.	
Capacity	236 c.c.
Unladen kerb weight ...	6¾ cwt.
Fuel consumption, driven hard	50.4 m.p.g.
Maximum speed	44.9 m.p.h.
Maximum top gear gradient	1 in 35
Acceleration:	
10-30 m.p.h. in 3rd	13.7 sec.
0-40 m.p.h. through gears	35.8 sec.
Gearing: 11.1 m.p.h. in top at 1,000 r.p.m.; 64 m.p.h. at 2,500 ft. per min. piston speed.	

UNIVERSALLY JOINTED at its base, the steering column tilts forward when the spring-counterbalanced door is opened, to provide unobstructed access to a bench seat 48in. wide. A pair of 12-volt, 30-watt headlamps are operated from a large battery which is charged from an A.C. generator and rectifier.

CONVENTIONAL CONTROLS of car pattern are used and a speedometer is mounted on the hinged bracket which supports the steering column from the front door. Visible in this picture are the two-spoke steering wheel (the test model has left-hand drive), pedals, gear lever (beyond the driver's left hand), ignition switch and lighting and direction-indicator switches.

petrol, even when driven hard. In Flanders we could not find really long and steep hills, but it was evident that this model (which originates within sight of the hot southern slopes of the Alps) is geared and cooled in a manner which envisages satisfactory go-anywhere performance. Acceleration from rest to 30 m.p.h. in 13.8 seconds, carrying an above-average load of two men and a heavy box of testing apparatus, means ability to get around town very briskly: although 35 m.p.h. is the most natural cruising speed, sustained full-throttle driving (meaning upwards of 50 m.p.h. downhill) worries neither the engine nor the chassis. Main road gradients slow the Isetta down very readily, but third gear, which provides a maximum speed of 34 m.p.h., allowed a speed of around 25 m.p.h. to be maintained comfortably on part throttle up quite considerable hills in the "Flemish Ardennes."

Despite its unorthodoxy, one of the charms of the Isetta is the paradoxical fact that in many vital respects it is perfectly orthodox. There is an ordinary bench seat which is spacious for two people and can carry three in emergency, the orthodox steering wheel and speedometer face the driver, the pedal controls are those of a normal car, and a diminutive gear lever beneath the driver's left hand moves in orthodox car fashion to select any of four synchromesh forward gears or the reverse ratio. To the driver and the passenger this is a small car, and should not be confused with vehicles which are more nearly enclosed motorcycles.

Conventional Standard of Suspension

On the example tested the seat was rather hard, but later production cars have thicker rubber cushions to give extra comfort. We cannot comment too precisely upon riding comfort, as the model loaned to us, although slightly modified from its original front suspension, was not quite to the latest specification, but the Dubonnet-type independent front suspension and telescopically-damped quarter-elliptic rear springs work very reasonably well. Belgium's atrociously rough cobblestones provide a severe test of a light car with small wheels, but they did not induce pitching or any more shake than many orthodox cars display. No reason for objecting to the short wheelbase or the forward seating position was evident.

Surprisingly, we found the steering rather heavy, although quite quick and very precise: heaviness seemed to result from friction, there being little self-centring action. Fast cornering reveals complete stability, with little body roll and quite a definite under-steer characteristic. Amply dimensioned hydraulic brakes tended towards fierceness, the rear wheels locking prematurely during our emergency stop tests and causing the car to swing sharply sideways—this was the only circumstance in which a driver became conscious of driving a short-wheelbase vehicle. Disappointingly, the turning circle is little better than on some conventional baby cars, the steering lock having perhaps been restricted to avoid the need for a differential in the narrow-track rear axle. The best lock is in the right-hand direction, most suitable for British traffic.

Light and Airy

All-round vision is excellent, the curved windscreen of toughened glass and the large side and rear windows of transparent plastic almost completely encircling the driver and passenger. Above, there is a canvas panel which folds back readily to provide fresh air and an upward view—hinged ventilation panels are used on each side of the body, and a single wiper blade clears a large area of the windscreen. At night, the small headlamps provide ample light, an alternator and rectifier charging the battery: included in the electrical system are parking lamps, stop lamps, and "winker" turn indicators.

A rich-mixture control is neatly paired

TRIANGULAR in shape, the Isetta chassis has independently-sprung front wheels, and a narrow-track rear axle with leaf springs and splayed telescopic shock-absorbers. Transmission is by all-synchromesh, four-speed and reverse gearbox, and enclosed duplex chain final drive.

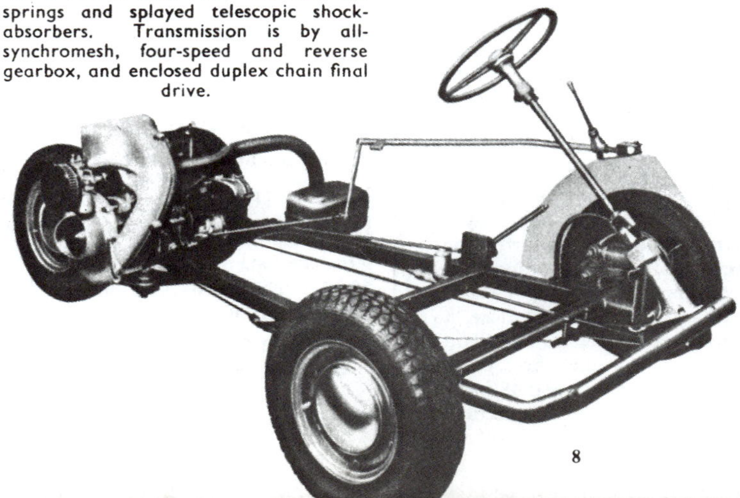

with the control for the electric starter, which latter is powerful enough to move the Isetta if it is accidentally left in gear. During our test the engine started readily whether it was cold or hot, and being air-cooled it very rapidly settled down to normal running temperature. Air is blown around the engine by a ducted centrifugal fan on the end of the crankshaft (there are no belts) and neither freezing nor overheating should ever be causes for worry.

The two-stroke engine, although sturdy, is perhaps the least attractive part of the vehicle. When idling it is none too smooth, a fast tick-over being necessary to overcome appreciable drag in the oil-immersed multi-plate clutch, and hard driving produces "two-stroke rattle" which sounds like pinking but is not eliminated by use of premium-grade fuel. Driven at moderate speeds, however, the engine is reasonably quiet and smooth, with no evident tendency towards the erratic four-stroking which mars the smoothness of many two-stroke engines during gentle driving.

Quite delightful is the four-speed gearbox, which is conventional save for the luxury of having simple and very effective synchromesh engagement of all ratios. The gears are quiet, and a reverse ratio is, of course, provided. The clutch seems very sensitive at first acquaintance, but soon proves acceptably smooth.

No luggage locker is provided but the whole space between the seat backrest and the rear window is occupied by a flat rubber-covered shelf, beneath which are the power unit and the rear wheels. A very considerable amount of luggage can be carried here, and provision for maps, etc., takes the form of a capacious pocket inside the door ahead of the passengers. Beneath the seat, there are a 12-volt battery and quite a comprehensive tool kit, whilst the very light spare wheel and tyre are behind the seat backrest, held securely in a fitted recess by a quick-release elastic strap.

Access to the power unit is through a panel in the right-hand side of the body. Unlike most two-stroke engines, this one does not require oil mixed with its petrol, there being a separate oil tank of ½-gallon capacity from which lubricant is pumped to the engine at a rate which varies with speed and throttle opening—the consumption is estimated at over 2,000 miles per gallon, there being no sump to drain and refill periodically. No fuel level gauge is fitted, but there is a reserve tap accessible from inside the car.

RUBBER MOUNTED on the right-hand side of the car, the "split-single" two-stroke engine is air-cooled by a fan mounted on its crankshaft. Electrical starting is provided, lubrication is by mechanical pump from a separate oil tank, and there is a petrol reserve tap.

A New Fashion ?

Boldly and cleverly unorthodox, simple but by no means crude, diminutive but quite good looking, the Isetta may well prove to be pioneering a development as important as that of the motor scooter which, also first popularized in Italy, has now become established throughout Europe and even farther afield. Entirely adequate for most short-distance journeys, and perfectly practicable for longer holiday-time journeys, the Isetta has manifest advantages in convenience and weather protection over most alternative types of ultra-light 3- and 4-wheeled vehicles. Very obviously, congestion in towns would be eased if a substantial number of orthodox cars could be replaced by vehicles as compact as this, and it may be noted that where orthodox vehicles are parked in a line alongside the kerb the Isetta may often be nosed straight into a gap (its length is less than the width of many commercial vehicles, and comparable with the width of an American car), allowing the passengers to step straight out of the front door on to the pavement.

PARKING is simplified by the shortness of the Isetta, here seen in proximity to a British 10 h.p. car. Nosed into the kerb, the Isetta gives direct access to the pavement through its front door.

Mechanical Specification

Engine

Cylinders	2, with common combustion chamber
Bore	48 mm.
Stroke: Front piston	64.3 mm.
Rear piston	66 mm.
Cubic capacity	236 c.c.
Piston area	5.6 sq. in.
Valves	Nil (two-stroke)
Cooling	Air ducted from fan on crankshaft
Compression ratio	6.5/1
Max. power	9.5 b.h.p.
at	4,500 r.p.m.
Piston speed at max b.h.p.	1,950 ft. per min.
Carburetter	Dell'Orto UC24B
Ignition	12-volt coil
Sparking plug	Marelli CW240 (14 mm.)
Fuel pump	Gravity feed
Lubrication	Total loss type, by variable-stroke mechanical pump automatically controlled from throttle.

Transmission

Clutch	Multi-plate, oil immersed, running at 0.46 of engine speed
Top gear (s/m)	5.17
3rd gear (s/m)	7.98
2nd gear (s/m)	13.15
1st gear (s/m)	22.86
Propeller shaft	Tubular with 2 universal joints
Final drive	Enclosed duplex chain
Top gear m.p.h. at 1,000 r.p.m.	11.1
Top gear m.p.h. at 1,000 ft./min. piston speed	25.6

Chassis

Brakes	Hydraulic (2 front brakes, 1 rear brake on solid axle)
Brake drum diameter	7 in.
Suspension:	
Front	Dubonnet-type I.F.S., with enclosed rubber-in-compression spring units
Rear	Quarter-elliptic
Shock absorbers:	
Front	Incorporated with springs
Rear	Telescopic, splay-mounted
Tyres	4.50—10

Steering

Steering gear	Worm
Turning circle: Left 32 feet, Right 24 feet	
Turns of steering wheel, lock to lock	2¼

Performance factors (at laden weight as tested):

Piston area, sq. in. per ton	10.9
Specific displacement, litres per ton mile	1,250 x 2 (2-stroke)

Coachwork and Equipment

Bumper height with car unladen:
Front (max.) 14½ in., (min.) 10½ in.
Rear (max.) 15½ in., (min.) 11½ in.

Starting handle	Nil
Battery mounting	Under seat
Jack	Lazy-tongs pattern
Jacking points	Front cross-member and rear axle
Standard took kit	Tool bag, jack and handle, 2 box spanners, 2 double-ended spanners, pliers, 2 screwdrivers, wire spark-plug brush.
Exterior lights	2 dipping headlamps, 2 side/direction indicator lamps, 2 tail/stop/direction indicator lamps, number plate lamp.
Direction indicators	Flashing type, non self-cancelling
Windscreen wipers	Single-blade electrical, non self-parking
Sun vizors	Nil
Instruments	Speedometer, distance recorder (without decimals or trip)
Warning lights	Ignition, direction indicators
Locks: With ignition key	Ignition
With other key	Door
Glove lockers	Nil
Map pockets	1 inside door
Parcel shelves	1 large shelf behind seat
Ashtrays	Nil
Cigar lighters	Nil
Interior lights	Nil (instrument light shines below facia panel)
Interior heater	Nil
Car radio	Nil
Extras available	Nil
Upholstery material	Cloth
Floor covering	Rubber
Exterior colours standardized	In Italy, Ivory; in Belgium, numerous alternative colours available.
Alternative body styles	Nil

Das Motocoupé

Isetta

ein neuer

Fahrzeugtyp

Endlich der neue wirtschaftliche Fahrzeugtyp, der den Motorisierungs-Wünschen weitester Kreise in zeitgemäßer Form entspricht: das Motocoupé BMW Isetta, ein Fahrzeug, das die Vorzüge eines Autos bietet, ohne die erhöhten Kosten eines Wagens zu verursachen.

Der Ein- und Ausstieg? Bequemer geht es nicht mehr: beim Öffnen der schrägliegenden Fronttür dreht sich die Steuersäule automatisch nach links und man kann ungehindert, aufrecht stehend das Fahrzeug betreten oder verlassen, ja geradezu hineinspazieren. Eine weitere Annehmlichkeit ist die vollkommen lautlos schließende Tür. Ein Griff – und schon ist sie fahrtgerecht geschlossen und gegen Aufspringen zuverlässig gesichert.

Dieser bequeme Ein- und Ausstieg erweist seine einzigartigen Vorzüge auch beim Parken. Man fährt in Fahrtrichtung zur Bordschwelle in die Parklücke hinein, parkt das Fahrzeug in dieser Stellung und steigt nach vorn aus, unmittelbar auf den Gehsteig.

Die Anordnung der Sitze: selbstverständlich nebeneinander, und zwar in geräumiger Breite, wie sie ein mittelgroßes Auto bietet. Das bedeutet, daß zwei Erwachsene und ein Kind auf der breiten, verstellbaren Polsterbank bequem Platz finden. Eine Beeinträchtigung der Bewegungsfreiheit des Fahrers ist schon deshalb ausgeschlossen, weil sämtliche Hebel zur Bedienung raumsparend linksseitig angeordnet sind.

Und noch eins: das Motocoupé fährt selbstverständlich aus Gründen der Fahrsicherheit auf vier Rädern.

So stellt die BMW Isetta die Ideallösung des Fahrproblems für alle dar, die ein wendiges, flinkes Fahrzeug bei geringen Betriebskosten ohne Park- und Garagen-Schwierigkeiten benötigen. Das gilt für Lohn- und Gehaltsempfänger zum Erreichen der Arbeitsstätte, für Ärzte, Vertreter, Handwerker, für den Einsatz von Reisenden großer Firmen, für selbständig Schaffende – und nicht zuletzt für die Damen, wenn der Hausherr mit dem »großen Wagen« unterwegs ist. Gegenüber diesen vielfältigen Vorzügen fallen die geringfügigen Betriebskosten, die überdies zum großen oder überwiegenden Teil steuerlich abgesetzt werden können, kaum ins Gewicht.

Der stabile Stahlrohrrahmen

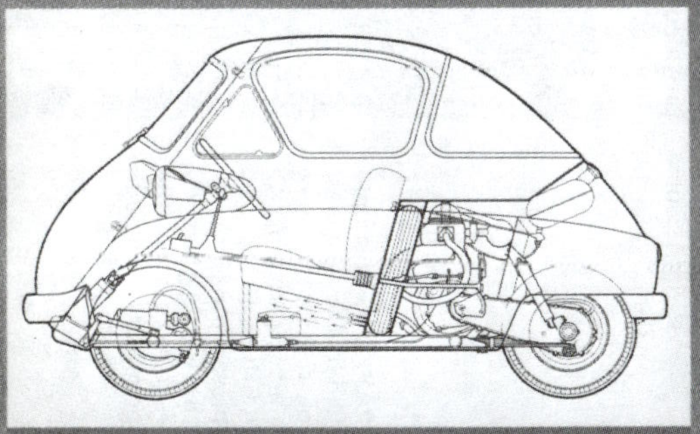

Motor	B M W Einzylinder Viertakt-Motor mit Gebläsekühlung
Bohrung	68 mm
Hub	68 mm
Hubraum	245 ccm
Verdichtung	6,8 : 1
Motorleistung	12 PS
Ventile	V-förmig, hängend im Zylinderkopf
Schmiersystem	Umlaufschmierung mit Zahnradpumpe
Kupplung	Einscheiben-Trockenkupplung
Motoreinbau	an rechter Wagenseite, hinter Sitzbank querliegend
Anlasser	12 V Dynastarter
Elektr. Anlage	12 V 130-W-Lichtmaschine, spannungregulierend
Getriebe	B M W 4 Gang-Getriebe mit Rückwärtsgang, am Motor angeflanscht
Hinterachsantrieb	über querliegende gummigefederte Gelenkwelle und kräftige, im Ölbad laufende, nachstellbare Duplex-Kette
Rahmen	stabile Stahlrohrkonstruktion
Achskonstruktion und Federung	Vorderräder einzeln an Schwingarmen mittels Schraubenfeder und Stoßdämpfer abgefedert Hinterräder an schmalspuriger Starr-Achse durch zwei Viertelelliptik-Blattfedern und Teleskop-Stoßdämpfer abgefedert

Räder	Stahlscheibenräder mit geteilter Felge zur erleichterten Reifenmontage		
Bereifung	5 fach, in der überdimensionierten Größe 4,80×10"		
Lenkung	Lenkrad mit untersetzendem Lenkgetriebe. Kleinster Wendekreis ca. 8 m		
Bremsen	Öldruckbremse, auf alle 4 Räder wirkend. Bremstrommel ⌀ 180 mm, Gesamtbremsfläche 320 cm²		
Hauptabmessungen			
Spurweite vorn	1200 mm	Länge über alles	2285 mm
Spurweite hinten	520 mm	Breite über alles	1380 mm
Radstand	1500 mm	Höhe über alles unbelastet	1335 mm
Gewicht			
Leergewicht	fahrfertig ca. 350 kg		
Nutzlast	230 kg		
Fahrleistungen			
Höchstgeschwindigkeit	85 km/st		
Steigvermögen	im 1. Gang über 30 %		
Kraftstoffverbrauch	ca 3,3 l/100 km		

Änderung der Konstruktion und Ausstattung vorbehalten.

Eine bekannte Tatsache: Auf vier Rädern fährt man sicher. Bewußt sind die Konstrukteure von dieser Überlegung ausgegangen. Der kurze Radabstand der beiden Hinterräder macht ein Ausgleichsgetriebe entbehrlich. Dadurch wurden erhöhte Richtungsstabilität und vorzügliche Straßenlage erzielt. Die beiden starr miteinander verbundenen Hinterräder werden durch die volle Kraftübertragung des Motors unmittelbar angetrieben. Feld- und Waldwege überwindet die BMW Isetta nahezu wie ein Geländefahrzeug.

Durch Viertelelliptik-Blattfedern mit zwei schräggestellten, hydraulischen Stoßdämpfern sind die Hinterräder weich gefedert, die Vorderräder sind einzeln aufgehängt. Die gute Bodenhaftung und Fahrsicherheit werden durch besonders griffige Reifenprofile der Laufräder noch erhöht. Beim Kurvenfahren laufen die Vorder- und Hinterräder in einem annähernd gleichen Radius. Die elegante Karosserie ist durch Rohre verstärkt und wird von einem stabilen Rohrrahmen getragen.

Die Breite der verstellbaren Sitzbank beträgt 1,18 m, der Abstand Sitzbankrand bis Vorderfront 0,70 m.

Alles in allem ein Fahrkomfort auf zwar wirtschaftlich bemessenem Raum, doch in einem gefälligen Fahrzeug, das in seiner modernen Form ausgesprochen flott und sportlich wirkt. Vor allem sitzen Sie sicher und wettergeschützt — selbst in bester Garderobe — in dem geschlossenen, geräumigen Motocoupé und haben ideal freie Sicht nach allen Seiten durch sichere Plexiglasfenster. Wird es im Sommer zu warm und Sie wollen während der Fahrt Luft und Sonne genießen, öffnen Sie mit einem Griff das Sonnendach sowie die beiden zugfrei lüftenden Ausstellfenster. Wird es im Winter zu kühl, so sorgt das Warmluftgebläse des Motors für wärmende Behaglichkeit. Hinter der Rücklehne befindet sich das Reserverad. Ebenso ist ausreichender Platz für die Gepäckaufnahme vorhanden. Die volle Fahrsicherheit bei Nacht ist durch die hochliegenden, lichtstarken Scheinwerfer gewährleistet.

Und nun beurteilen Sie die BMW Isetta einmal von dem Platz aus, der allein ein Urteil ermöglicht: hinter dem Steuerrad. Drücken Sie auf den elektrischen Anlasser... sofort springt der leistungsstarke, in aller Welt vieltausendfach bewährte BMW Viertaktmotor an. Er ist luftgekühlt, robust, zuverlässig und bekannt für seine Startfreudigkeit selbst im strengsten Winter. Leicht lassen sich die Gänge schalten. Das Viergangetriebe mit Klauenschaltung und Rückwärtsgang bietet weitgehend die Vorteile eines synchronisierten Getriebes. Die günstige Abstufung der vier Vorwärtsgänge ermöglicht rasche Beschleunigung. Beruhigt können Sie die volle Geschwindigkeit von über 85 km/st fahren, denn die hydraulische Vierrad-Öldruckbremse gewährleistet promptes und verläßliches Bremsen. Die Leistung und Bergfreudigkeit — Steigungen bis zu 32% werden mühelos überwunden — werden auch Sie überzeugen, denn der 250 ccm BMW Motor hat sich seit vielen Jahren unter allen nur vorstellbaren Fahrbedingungen hervorragend bewährt. Und es ist gewiß ein weiterer Vorteil, daß der BMW Kundendienst Sie, wie im Inland, fast in der ganzen Welt mit seinem werkgeschulten Fachpersonal hilfreich betreut.

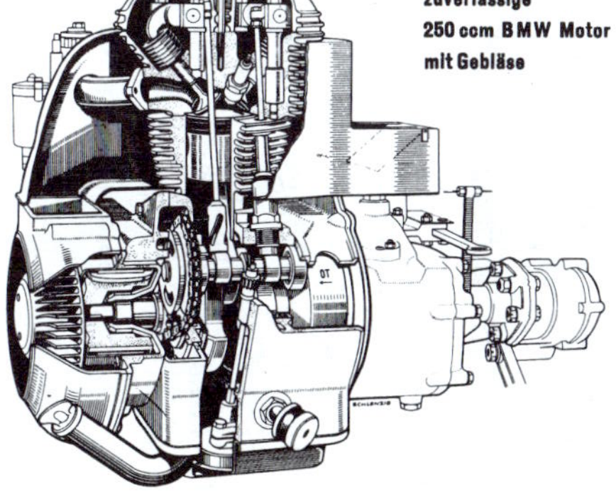

Der leistungsstarke, zuverlässige 250 ccm BMW Motor mit Gebläse

Daß es sich um ein bewährtes und zuverlässiges Fahrzeug handelt, hat der Erfolg des Motocoupés Isetta auf der Mille Miglia 1954 bewiesen. Auf der über 1600 km langen Strecke wurde der bemerkenswerte Durchschnitt von über 70 km/st erreicht. Eine Leistung, die in der ganzen Welt größte Beachtung fand.

Driving around with Walt Woron

Minimum transportation in the form of an Italian minicar, the Isetta . . . preview of Plymouth's new V8 engine in a Dodge truck, first impressions of the '55 Studebaker President, and tests on several cars equipped with Traction-Masters

Photos by Bob D'Olivo

Dropping body side-panel reveals minute powerplant. Air cleaner covers carburetor, hand points to single spark plug

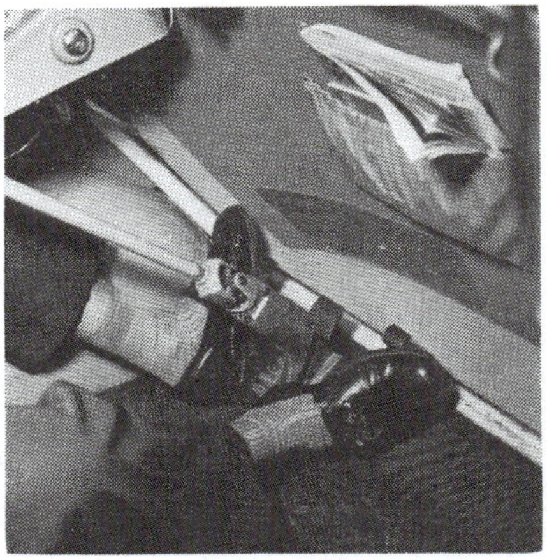

There's plenty of legroom in the Isetta, and surprisingly (for a small car), pedal placement allows unrestricted footwork

Removable seat folds for access to spare tire and gas tank's fuel reserve switch. Packages are carried on shelf above tire

The Isetta looks like a three-wheeled car because of narrow rear tread. Hydraulic brakes connect to only three wheels

ISETTA

Do you like a few laughs? Do you mind other people laughing *at* you, as well as *with* you? Do you like to stop traffic? Do you mind keeping alert *all* the while you're driving? Would you sacrifice acceleration and cruising speed for lots of gas mileage? Do you need a second car for the wife—just adequate transportation for taking her to the store? Well, if you've answered "yes" to all of these, then you might be a potential customer for the Isetta, product of the Iso Works, Milan, Italy. The Isetta is a car that was introduced to this country a few short weeks ago by an enterprising young man: Joe Haines (along with his partner, L. Novak). The enterprise comes from the fact that the Isetta is a minicar, and qualified at its factory price of $720 as one of the cheapest cars in the world. It's minimum transportation, and calling it a car would evoke considerable argument from some because of its tiny size (only 88½ inches long), its two-cylinder, air-cooled, valveless engine, its bench-type seat for two, its helicopterish body, all rolling along on four 4.50 x 10 tires that look like three.

The enterprise also comes from the fact that a minicar has never been acceptable to more than a scattered few in this country. (Remember the Bantam, the Austin, the Crosley, all minicars of sorts. Also, the Italian Fiat, the French Simca, the Renault?) Whether the Isetta can become more popular than these marques is dependent on how many people put low initial price, minimum cost of transportation, and smallness above everything else they ask of a car. At the same time they will have to put up with the feeling of a Lilliputian among a field of giants (if they are in any traffic at all), constant tension about maintaining a cruising speed anywhere near approaching that of surrounding traffic, the loud put-put of the exhaust, and constant shifting.

But—if you're looking for fuel economy, look no farther. That's the Isetta's forte. We've tested some "scotch" cars in the past, but none hold a candle to this one. In traffic driving I got an average of 37.9 mpg (tank mileage, not a controlled test), while on the highway I got 51.7 mpg! Overall I wound up with 44.5 mpg. Even if you drove this mite the annual average of 10,000 miles (which you probably wouldn't as a second car), can you imagine squawking about a yearly fuel bill that runs $64.26?! This is only about 60 per cent of the amount it would take you to run such a fantastically economical car as the Nash Metropolitan (basing both fuel bills on *overall* tank mileage, not the average of controlled tests as we did on page 20).

My first look at the Isetta was in a warehouse in Los Angeles Harbor. The car had just been swung off the ship from Italy. Getting it started was a familiarization course in itself: We found its fuel tank held 3½ gallons (three quarts in reserve), that it gravity feeds to a sleeve-type carburetor, that it gets its lubrication from oil mixing with gas, that it fires its single combustion chamber and

The ladies may not know the Isetta's wheelbase or weight, but they know it's easier to park and steer than other cars

Gearshift is to left and above wheel well. Only instrument is speedometer, mounted on panel connecting wheel to door

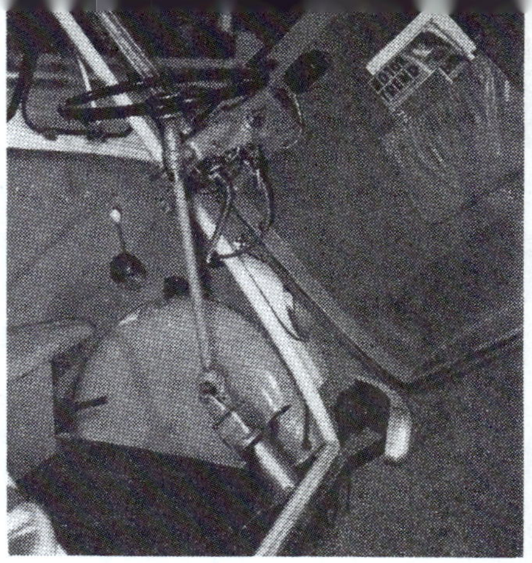

Parked head-on into a curb, you can step right out. As wide-swinging door swings forward, wheel swivels out too

works its two pistons with spark from one plug and a battery-fed coil, that it's easy to push (weighing only 665 pounds), that it has four speeds forward (and one reverse), an oil-bath clutch and chain drive to the rear axle (with no differential).

It took only a bit of fiddling with the carburetor to get it running properly and then the owner turned the car over to me for test. Surprisingly, the Isetta has ample legroom, headroom and shoulder room for two people; with three it's crowded. The seat-to-wheel relationship reminded me of a light truck (position-wise). The bench-type seat is fairly comfortable, but by no stretch of the imagination could it be termed plush or "deep-cushioned." There's vision aplenty; in fact, it's like looking out of a fish bowl.

Its short wheelbase (58½ inches) makes it pitch over practically any dip; fortunately it doesn't bobble around a lot after taking the dip, so the effect is only momentary. Body vibration is hardly noticeable at speed; at idle you'll feel it. Body sway around any turn is practically nil either to driver or passenger.

The Isetta is a pleasure to steer and has an easy turning wheel (no weight on the front end). It's a cinch to park. Its overall length is much less than the wheelbase of standard stock cars and equal to or slightly less than small sports cars. If you want, you can park it in half the space an ordinary car takes by heading directly into the curb. You then step right out onto the sidewalk.

The Isetta is definitely a car for around-town (I know, I tried it on a 200-mile highway trip!), but stay off of streetcar tracks. The front tread is about equal to the tracks, and rough spots in the asphalt pitch you aside; if you straddle the tracks, the two rear wheels get caught in the rough spots on each side of the track.

Acceleration is hardly noteworthy. Just for kicks, though, I checked it from 0 to 30 mph. I got a time of 13.8 seconds through the gears (first, second, third) which is about the time it takes a good, fast Hudson Hornet to hit 60. Other acceleration figures were: 10 to 30 in 11.8 seconds using third gear, 11.9 using second and shifting to third; 30 to 50 in 33 seconds using fourth gear.

A comfortable cruising speed with the Isetta is between 35 and 40 mph. Maximum cruising speed (at which you're pushing it) is 40. Top speed is around 47 mph (at least with this fairly new one).

Joe Haines tells me that he's going to try to deliver Isettas at $975, Port of Entry. It appears to be about as low as he can profitably and reasonably price so unfamiliar a version of the auto.

DODGE FIREARROW

Hearing about a Dodge Firearrow (by Ghia) that was in town recently, I naturally requested an opportunity to drive it It was appearing in a Dodge showroom in Pasadena (Calif.) and was to spend a couple of days at another dealer's in Compton (about 25 miles away) before being shipped to the California State Fair in Sacramento.

Arrangements completed, I jumped into my car, drove to Pasadena, where I met Danny Eames, driver and engineer for Dodge. Danny explained to me that this version of the Firearrow was made in '53, and therefore had the '53 engine and Gyro-Torque transmission. I had hoped it would have been one of the hotter versions, like the 245-horsepower job Betty Skelton drove at 144 mph at the Chrysler Proving Ground Dedication, but still welcomed the chance at trying it out.

The car has more punch than stock, being lighter (3200 pounds vs. 3680 for our test Dodge), giving it a better power/weight ratio. Although I had no opportunity to time the Firearrow's acceleration, I noticed that it left most stock cars at all the signals. With Gyro-Torque, the Firearrow had an exceptionally smooth shift—a passenger wouldn't notice it.

Vision all around is very good: I could see the right front fender from deep in the cushioned straight-across seat. Three can sit abreast, but the person in the center would have to straddle the large tunnel. The wooden-rimmed, racing-type steering wheel felt somewhat smaller in diameter than most wheels, but still gave enough leverage for taking any corners without getting your arms all tangled up. The spokes, however, hid some of the instruments (placed in two circular dials). The constant-curve one-piece windshield reflected the dash more than one with straight sections and curves.

Broad, sharp-edged band of chrome is almost the only applied decoration on the Firearrow. It is perhaps the most successful meeting of European and U.S. designs

ROAD AUTO AGE TEST

It may not look much like a car, but this Italian midget is a remarkable machine indeed.

In a full power drift at 30 mph the tiny car shows practically no lean at all. It handles and steers like a true sports car.

meet the ISETTA

IF SOMEONE told you that you could buy a brand new sports car for only $1,000 that would seat three, get 70 miles to the gallon and weigh less than 700 lbs., you would probably think he was crazy. Well, technically you'd be wrong because there is such a car; it's called the Isetta.

Now it must be admitted that at first glance the tiny Isetta doesn't really look like a car at all, much less a sports car. But after hearing of its many achievements, looking it over carefully and then driving it, I came to the conclusion that the only thing you *can* call it is a sports car. Take, for instance, its record in competition. (That's what I said, Harry—competition.) In the 1954 Mille Miglja, one of the toughest road races in the world, Isettas finished one, two, three (Continued on next page)

Rear wheels are set so close together that from some angles the Isetta resembles a three-wheeler. Wheelbase is 58.5 in.

With the huge front door open it is a simple matter to climb in and out of the Isetta. There is lots of leg room and the visibility is without equal.

Meet the Isetta

on Index of Performance! Why? It's a handling fool. Doesn't look like it, though, does it? That's what I thought until I climbed—I mean walked—behind the wheel. Here's what happened on my first ride:

Photographer Dan Rubin and myself went to pick up our test car from Tony Pompeo, a well-known sports-car dealer in New York City. Now Dan is just under six-foot three and fairly husky and while I am not particularly large I weigh in at a good 170, which is heavy for a midget. When we got our first look at the Isetta we both thought out loud, "Naw, couldn't be!" But it was. Here was a strange-looking but beautifully-made Italian car that looked as though it had been made to carry two or three small puppies and one medium-sized can of tomato soup and yet Tony was trying to tell us that the two of us were actually going to fit into it. While we were laughing, Tony opened the one huge ice-box-type door in front of the car, turned the key, pulled the starter and instantly the little two-cylinder, two-cycle motorcycle engine screamed bravely and loudly at us. The exhaust pipe was on the side, I noted. When the car had warmed up Tony asked us to step in and when we did we were both amazed. With the door shut there was actually more than enough leg room and plenty of space on the bench seat for both of us. What was even more unusual for a small car, the floor pedals were completely unobstructed and since the engine is in the rear there was not even a transmission hump.

After being checked out on the operation of the various controls, Dan and I headed for Central Park, which must have been laid out by a sports-car fan. Talk about being conspicuous! We couldn't have attracted more attention if we had walked down the street stark naked. Everyone—I mean *everyone*—stopped to look at the Isetta. They all seemed to be somewhat amused. Me? I was somewhat amazed at all the acceleration this little car had, considering its 236-cc, 9.5-hp engine.

Before we had gotten into the park I had noticed something else, too. Despite its size and low price the Isetta has a four-speed transmission that's a doozie. There's no foolishness like synchromesh, of course, but once you get used to it, which takes only a few short minutes, you find that full power shifts are the order of the day either going up or coming down through the gears. And no matter how you try, the gears just won't clash. In fact, I don't think I've ever handled a nicer gearbox on any car, regardless of price. That is one of the major secrets of the Isetta's success.

But there are others. As we drove through the winding roads of Central Park all of the drivers around me were blowing their horns and laughing like idiots and soon I'd had just about enough. So I put my foot to the floor and let 'er rip. Tony had told us that the Isetta's top speed, at the end of a long run, was about 60 mph and I know I never had the car going that fast. But what I do know is that with the Isetta running wide open, going around curves at full speed, we stayed with and beat most of the cars in sight.

After three of four fast "laps" around the park, we decided to make a few acceleration tests, just for fun. Zero to 30 mph came out to just over 11 seconds and the car took another 11 seconds to reach 40. Still, probably because of the car's size, there was a definite feeling of punch and responsiveness. And, if you wind the engine out to 4,500 rpm or so in each gear, you get up to a respectable cruising speed in surprisingly little time. A speed of 50 mph may not seem like much, but with a vehicle that handles like this one you can put up an average pretty close to your cruising speed. There's just no need to slow down for corners.

When you do decide to slow down, however, you're in for another shock. The brakes on the Isetta can only be described as incredible. Say you are touring along at 30 mph and the traffic light half a block away changes to red. If you should want to show off in such a situation all you do is wait until you are about 20 ft. from the intersection and then tromp on the brake pedal. You will stop —not quickly, but immediately. There is no nose-diving or hesitation. There is just an audible thud as the brakes take hold—then silence, almost as if you had run into something. It's quite a sensation. Naturally you can make nice, smooth stops, too. But what fun is that?

As far as gas mileage is concerned, I wouldn't swear to that 70-mpg figure because I didn't keep the car long enough to do an accurate test myself, but the car really should do at least that well considering its low weight and extremely small engine. When you realize that the Austin A-40, for instance, will get up to 40 mpg using a power plant approximately five times the size of the one in the Isetta, 70 mpg doesn't seem impossible at all. And I mean for an average—not maximum mileage.

I know what you are thinking, of course; is the Isetta really practical at all? In answer let me say that if you don't mind being stared at as though you had three heads, and if you could use a second car for light shopping or going to and from the station, the Isetta may be worth looking into. The visibility is as good as it would be in any fishbowl and there is even a "convertible" top that swings back like a sort of fabric sun roof for use in warm weather. Instrumentation is as simple as possible—the only honest-to-goodness gauge is a speedometer—but an electric windshield wiper, left-right indicator and two sets of headlights are standard equipment. There is even an ignition warning light. The gas tank holds about three and a half gallons and you can judge whether or not it is full by the shake-and-listen method. If you decide to use the car for picnicking, the entire bench seat can be removed and placed under your favorite tree. A spare tire is stored behind the back of the seat. As I said, the Isetta's major points are great handling and unbeatable economy and if your taste happens to run to animated kiddie cars, this may be just your dish. Remember, it only costs $995.●

SPECIFICATIONS

ENGINE: two-cylinder, two stroke, with single spark plug; displacement, 236 cc; developed hp, 9.5 at 4,500 rpm; compression ratio, 6.5 to 1; single carburetor; gravity fuel feed; forced air cooling; multiple-plate oil bath clutch.

TRANSMISSION: four forward speeds and reverse (crash box).

REAR AXLE: one block with rear drive gearbox containing double-roller chain in an oil bath.

SUSPENSION: front wheels individually suspended on heavy rubber pads with friction-type shock absorbers; rear wheels equipped with laminated springs and hydraulic shock absorbers.

DIMENSIONS: wheelbase, 58.5 in.; tread, 45.8 in. rear, 19.5; over-all width, 52.3 in.; over-all length, 12 ft., 3¾ in.; height, 51.5 in.; weight, 665 lbs. (curb weight).

PERFORMANCE

ACCELERATION:	Zero to 30 mph in 11.2 seconds
	Zero to 40 mph in 22.4 seconds
TOP SPEED:	60 mph (approx.)

DIMINUTIVE MOTOCOUPE BY BMW MAY BE ANSWER TO COMMUTERS' LOCAL TRANSPORTATION PROBLEMS

Note location of gear lever left; spare tire and storage space behind driver. No space has been wasted.

ISETTA SUBURBANITE

You don't have to be a midget to ride comfortably in the Isetta. There is ample leg room even for a six footer and the seat is adjustable backward and forward. The Isetta is as long as a Cadillac is wide, with small turning circle.

In addition to the ventipanes which sweep air in and out at speed the canvas top can also be folded back on sunny days to enjoy the view above. In rainy weather the car seals nicely and visibility front and rear is extremely good.

If you are a suburbanite who needs a second means of conveyance that is suitable for use rain or shine, hot or cold, $800 will get you the Isetta Motocoupe which was introduced in April in Germany and may soon be imported to the United States. Designed originally in Italy, but developed and perfected by BMW (Bayerische Motoren Werke AG) Munich, Germany, the Isetta has basic construction characteristics of a completely new type vehicle.

It would be a mistake to regard it as a simplified version of a regular car or as a type of scooter. Therefore the word "motocoupe" was especially created for it. The outstanding feature of B.M.W.'s Isetta Motocoupe is of course the fact that it presents a new solution to the traffic problem as it is only as long as a Cadillac is wide.

There is only one door, located in front, with the steering wheel attached to it. The steering wheel automatically swings to the left as the door opens and the driver can get in or out of the car standing practically upright. The door closes soundlessly and locks airtight with a safety lock.

Safety was the main consideration in construction the B.M.W. Isetta. The four wheels are attached to a non-twisting tubular frame which makes a stable base for the body. The short distance between the rear wheels eliminates the necessity of a differential and increases the car's directional stability and makes for superior roadability. The locked rear wheels are directly driven. Quarter-elliptic leafsprings with two diagonally located shock absorbers provide soft springing for the rear wheels. The front wheels are individually suspended.

The folding roof, as well as the no-draft side windows open with a twist of the wrist. Two adults and one child can sit comfortably on a backward and forward adjustable benchtype seat, 46 inches wide (compared to 52 inches in a standard American car). A distance of 28 inches to the door allows for ample leg room. To give additional space, the gearshift is on the left and all the other controls are grouped around the steering wheel. The B.M.W. Isetta Motocoupe has generous space for parcels on a shelf in rear of the driver and the spare wheel is located behind the seat

The motor is equipped with a radial aircooling blower. The displacement is 250cc and the horsepower is rated at 12. The motor has proven itself successful as a power plant for the one cylinder B.M.W. motorcycle all over the world. The rugged, aircooled pow-

ISETTA SUBURBANITE

Chassis of the Motocoupe is tubular construction to provide exceptional rigidity. Brakes are hydraulically actuated and apply to all four wheels. Since the track in the rear is only 25 inches no differential is used. Note clutch, brake and accelerator pedals at steering wheel post.

Below: the engine is a one cylinder, four cycle job with hemispherical combustion chambers and overhead valves actuated by push rods. Blower provides additional cooling and also heats the passenger compartment in cold weather.

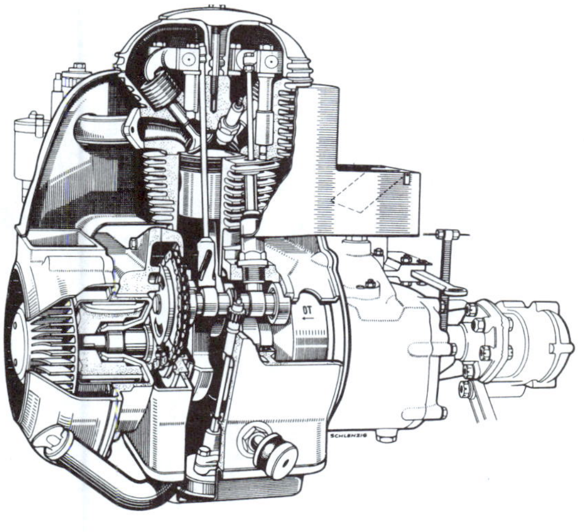

As the front entrance door is opened the steering wheel swings to the left providing free exit or entrance. Headlamps are a little below normal height but directional signals also are provided. Seat is 46 inches wide which is roomy enough for two adults and a child.

erplant, connected to a 5 speed transmission (four forward, one reverse) gives surprising pick-up and flexibility in traffic. A 32% uphill grade is easily negotiated by the B.M.W. Isetta.

The motocoupe consumes one gallon for 60 miles. Top speed which slightly exceeds 50 miles an hour can be safely reached as the vehicle is equipped with hydraulically operated brakes. The smoothness of the motor at idling and at higher speeds should be emphasized.

The B.M.W. Isetta Motocoupe is a commuter's dream and ideal for anyone who appreciates a flexible, comparatively fast vehicle which negligible upkeep and virtually no parking and garage problems

Albrecht Goertz, former design consultant for Motorsport and now design consultant for BMW, brought an Isetta to the Motorsport office early in June. Members of the staff were given an opportunity to ride in and drive the little bug which handled well and held its own in the maze of fast moving traffic on Fourth Ave., New York. The demonstration was quite impressive but the motocoupe's chief forte, of course, would be in the suburbs for traveling to and from the station, for short hauls to stores and schools and just for fun in driving about town with no parking problems to fret about.

The front thread is 47.5 inches and the rear 25 inches. Wheelbase is 59 inches and overall length 90 inches, over-all width 54⅜ inches. Weight is about 700 lbs. with payload capacity of 460 lbs. The single cylinder engine is a four cycle with a bore of 68 mm and stroke of 68 mm giving a displacement of 245 cc. The combustion chambers are hemispherical with pushrod operated overhead valves. Compression ration is 6.8 to 1. Tires are 4.80 by 10.

If the Isetta is imported in line present planning it is likely that it will be offered more-or-less as a mail order deal with no established sales outlets as such On the other hand, the Isetta may find a good outlet through bicycle dealers. No serious service problems are anticipated.

FRONT acts as door, gives easy entry to simply finished interior.

NO, it's not a mechanised egg. It's the Italian-built Isetta two-seater—a very practical, efficient little vehicle, despite its curious shape.

First of its brood to reach Australia, it was shown at the recent Italian Exhibition in Sydney, where its agents, Halifax Motors, collected a lot of orders at £555 a pop, including tax.

The Isetta is made in Milan by the makers of Iso motor-cycles; it is the firm's first venture into the car field.

Designed for cheap motoring, easy parking and manoeuvrability in city traffic, the car has a top speed of 58 m.p.h., and will do 82 miles per gallon of petrol, plus one-

Nippy performance, surprising comfort make this tiny newcomer a very practical city car

Eye-opening Isetta

fifth of a pint of oil (added to the petrol to provide lubrication).

The whole front of the car is one big door, to which the dash panel is attached. The universal-jointed steering column swings out of the way when the door is opened.

Visibility is remarkable. Almost the whole top of the car is transparent, strong plastic being used for all the windows. A soft plastic sheet on the roof rolls back to let in extra air or sunshine, and two small side-window panels also swing out for ventilation or for hand-signalling (though the latter may be a little awkward).

The body is of light aluminium panelling, built on to a floor of 21-gauge steel. The whole structure is reinforced by a strong tubular steel frame.

The tiny twin-cylinder, two-stroke engine develops 9.5 b.h.p. at 4500 r.p.m. Bore is 48 mm., stroke 60 mm., total capacity 236 c.c. Compression ratio is 6.5 to 1. Coil ignition, forced-air cooling, and variable-stroke oil pump are other features of the engine.

The high power-weight ratio of 65lb. per h.p. should give the Isetta excellent hill-climbing and acceleration abilities. Specifications suggest the car will climb a 1 in 25 grade in top gear, and a 1 in 4 grade in first.

The gear lever, which is mounted in the side of the cockpit next to the driver, operates a four-speed box which allows 14 m.p.h. in first gear, 23 in second, 38 in third, and 58 in top.

A multiple-plate wet clutch is used, and final drive is via a double-roller chain in an oil-bath.

The car is only 90in. long, 53½in. wide, and 53in. high. Wheelbase is 60in., kerb weight 6 cwt. Track is uneven, being 48in. in front and only 20in. at rear. The low-pressure tyres are 4.50 by 10 inches.

The petrol tank holds 3¾ gallons, of which ¾ of a gallon is held in reserve and is obtained by turning the special tap. Petrol is fed to the carburettor by gravity. Oil capacity is 5½ pints.

Front suspension is independent, with friction-type dampers; laminated springs with hydraulic shock absorbers are used at rear.

Hydraulic brakes acting on 7¼in. drums on all four wheels are supplemented by a mechanical linkage to the rear wheels for the hand-brake.

Despite its small size the car holds two very comfortably. There is a fair amount of luggage space in a well behind the seat and on top of the spare wheel, lying on a shelf behind the well.

The very respectable performance of the Isetta is pointed up by its showing in the Mille Miglia (Italian 1000-mile road race) this year. Four Isettas started; all finished, and the fastest one was reported to have averaged nearly 50 m.p.h.

Another indication of success is the fact that the car is now being built under licence in Germany (by B.M.W.) and in France (by Talbot). Its makers freely admit that although the Isetta is ruggedly built, it should not be used consistently on rough roads. But that is not its purpose—it is essentially a city car, and for this function it seems to be ideally suited. • • •

***TINY DASH** is seen through steering wheel; clutch, brake and accelerator pedals are at foot of column, gearshift in side panel.*

***TWO-STROKE** engine is concealed behind a louvered flap in left side of body. Of 236 c.c., it gives Isetta a top speed of 58 m.p.h.*

***REAR VIEW** shows huge expanse of plexiglass windows, giving true all-round visibility. Rear track is very narrow—only 20 inches.*

Although of unusual shape, the little Isetta is functional. A large area of glass and plastic provides good visibility (although the windscreen pillars could be reduced in width with advantage) and the canvas roof panel can be opened easily. Winking indicators are fitted on the sides. The engine cover can be seen forward of the narrowly tracked rear wheels

The Autocar ROAD TESTS

No. 1581 B.M.W. ISETTA MOTOCOUPÉ

Left: An exterior mirror is fitted because the view through the rear window at times may be obstructed by luggage on the rear shelf. There is only one windscreen wiper blade. The head lamp beams are such that the car can be driven at maximum speed at night with complete confidence

AMONG a number of unusual approaches to the difficult task of producing a really economical small car, one of the most interesting is the B.M.W. Isetta Motocoupé, which is now available for general sale in Great Britain. With its single door at the front and its unusual shape, it attracted considerable interest at the recent show at Earls Court.

Before a description is given of the capability of this model it is worth mentioning some of the difficulties that beset the designers of a car of this type. It is important that such a vehicle should feel like a "real car," and perform at least well enough to keep up with the speed of an average traffic stream; yet if it is to be really economical it must have a small engine and be of light weight. A high initial price tends to defeat the object of the car, and small firms who have entered this field have been faced with the problem of not being able to use many components common to cars produced in large numbers. Door handles, for example, in the range available from component manufacturers are too big and heavy, but to have them made specially makes the car expensive. Consequently, entry into the economy field by a firm with resources for high output—and thus reduced cost—is welcome.

Originally the Isetta design was produced and developed by Iso in Italy, and that company is still producing a few cars. The Iso models attracted attention by their splendid performances in the Mille Miglia, the thousand miles race in Italy. Some cars are also being made in France, but as B.M.W. have acquired the manufacturing rights, they are expected in time to be the sole makers, and it is the B.M.W. cars that are sold (through A.F.N., Ltd.) in this country.

The Isetta has a box-section frame, the spacing of the side members narrowing towards the rear because the back wheels are much closer together than those at the front. At the front there is a strong tubular extension, and another sturdy tube, following the body profile, forms the opening of the door. Although the kerb weight is kept down to little more than 7cwt, there is nothing flimsy about the car. It feels robust, although, of course, the door panel itself could be expected to provide little protection for the occupants in the event of an accident. The handling is not affected to any great extent by the pronounced difference between front and rear track.

The engine starts at a touch of the electric self-starter in all conditions and, being air cooled, it warms up quickly. However, in common with other single-cylinder units, the slow running leaves something to be desired. When in perfect adjustment the engine will probably provide satisfactory slow running, but on the car tested it was necessary to use the throttle pedal to keep the engine turning during traffic

The neat 245 c.c. single-cylinder engine is not very accessible, but the carburettor and oil filler-cum-dipstick can be reached easily. The rotary fins for air cooling are also visible

No. 1581 B.M.W. ISETTA . . .

hold-ups. This would be of small account, were it not for the unhappy placing of the brake lever under the driver's left thigh. To grasp this lever while the engine is kept running with the throttle pedal at traffic lights is not comfortable.

On such a car the ease of control is of special importance, and it is pleasing to find that the Isetta is responsive. The unusual design does not result in the car being difficult to handle. Even at its maximum speed of something over 50 m.p.h. it remains stable and will corner with little roll. Understandably acceleration is inferior to that of most cars, but if full use is made of the gears and available power the car certainly does not hold up other traffic.

At first glance the Isetta seems intended purely for local use, but its potential speed and stability do make it practical for long runs. For example, when the car was driven hard for about 200 miles, of which more than half was on the congested Great North Road, it was passed only four times. This is not intended to suggest that much of the traffic on the road could not have passed the Isetta, but it does illustrate that in everyday traffic conditions the ground can be covered as quickly in this model as it is by most drivers of more powerful cars.

The small wheels and short wheelbase make it sensitive to poor road surfaces, and in some conditions the car pitches considerably, speed having to be reduced in the interests of comfort. After delivery to *The Autocar* the car was driven several times over a rather poor "commuters' road" and it was felt that a long journey would be tiring because of pitching. It was a pleasant surprise to find later that—partly owing to the generally excellent surfaces of British roads—long journeys produced little discomfort. There is a tendency for the driver's back to rub against the seat. This was found easy to overcome by the insertion of a thin pad of foam rubber.

The single door cannot be slammed in the ordinary way, the strong handle having to be turned to shut the door securely. This helps to hold the door firmly, so that there is no interference with the accuracy of the steering; the column is secured to the door, and hinges on a universal joint. The driving position is of the "sit up and beg" type. It is comfortable and, with the exception of the hand brake which has already been mentioned, all the controls are well placed. On the steering column are two levers controlling respectively the flashing indicators and the dip switch. The speedometer is directly in front of the driver. It could be read more easily if the dial were not mounted a little out of true in an anti-clockwise direction. The starter is operated by the ignition switch.

Left: The front of the car is the door. The steering column has a universal joint at its base, and the steering wheel moves with the door

The rear wheels are close together, so that the car looks like a three-wheeler from some angles

The clutch and gear box could be improved. Not only are they frequently in use, but the Isetta has an important second-car application—for taking children to school and for shopping. We would, therefore, have expected the designers to make sure of obtaining a smooth clutch take-up and easy gear-change. Once the Isetta is moving the clutch mechanism is satisfactory, but when starting from rest that on the car tested was fierce. Considerable skill was required to produce a really smooth getaway.

Although it is possible to start the car from rest in second

The canvas roof is secured by two catches, and can be opened in a moment. Just behind the seat is the petrol on-off and reserve control. A large number of parcels, or some luggage, can be accommodated on the shelf behind the seat. The rear and main side windows are of plastic

Apart from warning lights, the speedometer is the only instrument. This is marked in k.p.h., and distance recording is in kilometres. The gear lever is on the driver's left, and near it is the choke. The direction indicator control and dip switch protrude from the steering column

gear, first is the natural choice as the power is limited to that provided by a very small engine. This gear is difficult to find when the car is stationary. It can be so elusive that a driver usually adopts the habit of slipping into first gear at the moment of coming to rest at traffic lights and similar hold-ups. When the gear lever has been moved to neutral as the car slows for some obstruction, it is sometimes difficult also to find second gear if it is required while the car is still moving at any speed appropriate to that gear.

Braking efficiency is satisfactory. Moderately firm pedal pressure provides good retardation, which is smooth and evenly distributed among the wheels. Only on poor surfaces is the effect unsatisfactory, and the fault in these circumstances is a result more of the short wheelbase than of the brakes themselves. As the brakes are applied a quite violent pitching motion begins, but this occurrence is rare in normal driving.

Apart from the canvas roof panel, ventilation is provided by swivelling windows in the forward edges of the side windows. The one on the passenger side is effective, but the driver's window can be opened only to a very small extent because of its close proximity to the driver's knuckles as he holds the steering wheel.

The Isetta has a number of intriguing features. The front opening door makes people stop and stare, and even more sure of collecting a crowd is the technique for parking between two fairly large vehicles left close together. Provided that the gap between them is an inch or two more than the width of the Isetta, the little car can be driven to the kerb forwards, the driver and passenger stepping directly on to the pavement. When the door is open the steering is not affected, and in car park manœuvring, while the driver may be awaiting his passenger, considerable amusement is caused by first sight of the Isetta being turned while the door is open.

Selling at a total price in this country comparable with that of the least expensive orthodox cars, and providing a high degree of economy, this B.M.W. may well find a niche in the British market. There are no innovations incorporated in the design or construction that are not included for practical reasons; thus although the car is unusual it is honest.

B.M.W. ISETTA MOTOCOUPÉ

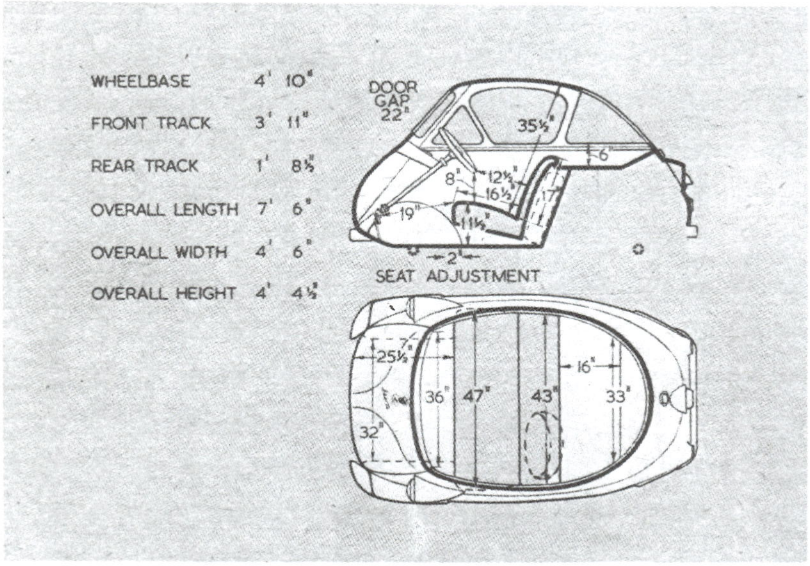

Measurements in these ¼in to 1ft scale body diagrams are taken with the driving seat in the central position of fore and aft adjustment and with the seat cushions uncompressed

PERFORMANCE

ACCELERATION: from constant speeds.
Speed Range, Gear Ratios and Time in sec.

M.P.H.	2.70 to 1	3.54 to 1	5.17 to 1	10.05 to 1
10—30	—	17.0	11.4	—
20—40	30.4	18.7	—	—

From rest through gears to:

M.P.H.	**sec.**
30	12.5
40	22.7

Standing quarter mile, 29.8 sec.

SPEEDS ON GEARS:

Gear		M.P.H. (normal and max.)	K.P.H. (normal and max.)
Top	(mean)	50.8	74.5
	(best)	54	79.2
3rd		39—41	57—60
2nd		22—30	32—44
1st		10—15	14.5—22

TRACTIVE RESISTANCE: 48.7 lb per ton at 10 M.P.H.

TRACTIVE EFFORT:

	Pull (lb per ton)	Equivalent Gradient
Top	71.6	1 in 32
Third	120.8	1 in 17.5
Second	208.3	1 in 10.8

BRAKES:

Efficiency	Pedal Pressure (lb)
28 per cent	25
39 per cent	50
69 per cent	75

FUEL CONSUMPTION:
61 m.p.g. overall for 158 miles. (4.6 litres per 100 km.)
Approximate normal range 52—73 m.p.g. (5.4—3.9 litres per 100 km.)
Fuel, Premium grade.

WEATHER:
Air temperature 51 deg F.
Acceleration figures are the means of severa runs in opposite directions.
Tractive effort and resistance obtained by Tapley meter.

SPEEDOMETER CORRECTION: M.P.H.

Car speedometer:	10	20	30	40	50	56
True speed:	10	19	31.5	38.2	47	54

DATA

PRICE (basic), with coupé body, **£291 17s 9d.**
British purchase tax, **£123 2s 3d.**
Total (in Great Britain), **£415.**
Extras: Radio **£22 1s 0d.**

ENGINE: Capacity: 245 c.c. (14.83 cu in).
Number of cylinders: 1.
Bore and stroke: 68 × 68 mm (2.67 × 2.67in).
Valve gear: o.h.v., pushrods.
Compression ratio: 6.8 to 1.
B.H.P.: 12 at 5,800 r.p.m. (B.H.P. per ton laden 25.3).
Torque: 25lb ft at 5,800 r.p.m.
M.P.H. per 1,000 r.p.m. on top gear, 21.

WEIGHT: (with 2 gals fuel), 7 cwt (784lb).
Weight distribution (per cent): F, 54.9; R, 45.1.
Laden as tested: 9½ cwt (1,064lb).
Lb per c.c. (laden): 4.3.

BRAKES: Method of operation: Hydraulic.
Drum dimensions: F, 7in diameter; 1¼in wide. R, 7in diameter; 1¼in wide.
Lining area: F, 25 sq in. R, 25 sq in. (105 sq in per ton laden).

TYRES: 4.80—10in.
Pressures (lb per sq in): F, 20; R, 18.

TANK CAPACITY: 2.2 Imperial gallons.
Oil sump, 2¼ pints.

TURNING CIRCLE: 24ft (L and R).
Steering wheel turns (lock to lock): 2¼.

DIMENSIONS: Wheelbase: 4ft 10in.
Track: F, 3ft 11in; R, 1ft 8½in.
Length (overall): 7ft 6in.
Height: 4ft 4½in.
Width: 4ft 6in.
Ground clearance: 6in.
Frontal area: 12 sq ft (approximately).

ELECTRICAL SYSTEM: 12-volt. 31 ampère-hour battery.
Head lights: Double dip; 25-25 watt bulbs.

SUSPENSION: Front, independent coil springs. Rear, quarter elliptic.

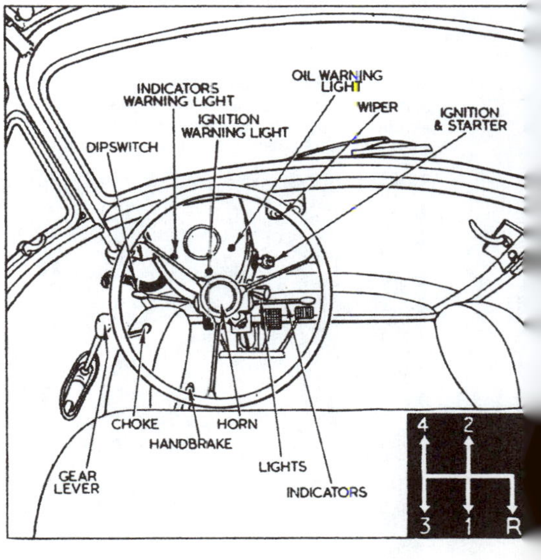

Isocarro 500

UTILIZZAZIONE TELAIO

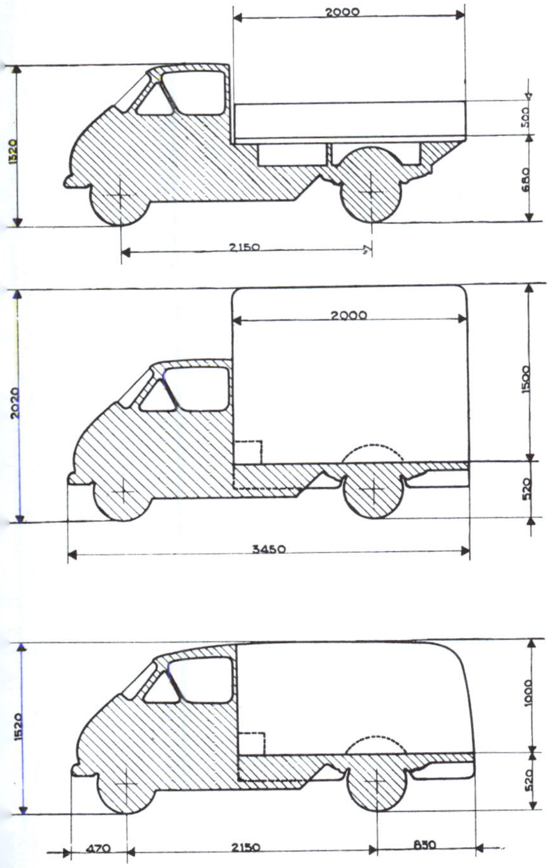

CARATTERISTICHE TECNICHE

MOTORE: ciclo a due tempi a distribuzione corretta, cilindro sdoppiato - alesaggio di ciascuna canna mm 48 - cilindrata totale cm^3 236 - potenza effettiva Cv 9,5 a 4500 giri. - *Alimentazione* a benzina per caduta, carburatore Dell'Orto. - *Lubrificazione* separata con pompa meccanica. - *Raffreddamento* forzato ad aria. - *Avviamento* elettrico. - *Accensione* a batteria con spinterogeno. — **FRIZIONE:** a dischi multipli in bagno d'olio. — **CAMBIO:** 4 velocità e retromarcia. Velocità max. a pieno carico km/h 65. — **ALBERO TRASMISSIONE:** costituito da un elemento tuborale e da due giunti elastici. — **PONTE POSTERIORE:** differenziale con coppia conica (Rapp. 12/40). — **TELAIO:** ad elementi tubolari prolungantisi fino oltre le ruote anteriori. — **SOSPENSIONI:** le anteriori indipendenti con molleggio a tamponi di gomma speciale. Sospensione posteriore a molle a balestra, con ammortizzatori differenziali idraulici. — **STERZO:** guida sinistra - scatola di riduzione a vite a madre vite, con grandissima superficie di contatti. — **FRENI:** idraulici sulle 4 ruote con tamburi di mm 180; meccanico sulle ruote posteriori. — **RUOTE:** a dischi scomponibili. — **PNEUMATICI:** 5.00-10. — **IMPIANTO ELETTRICO:** costituito da un alternatore autoregolatore ad alta frequenza, privo di collettore e spazzole. Tensione 12 Volt - batteria 24 Amp. - motorino d'avviamento - lampeggiatori - luci di posizione - spie - fanali - tergicristallo - avvisatore acustico.

DIMENSIONI - PESI (Ingombri con cassone normale aperto)

Passo	mm	**2150**	**Lunghezza**	mm	**3450**
Carreggiata anteriore	»	**1200**	**Larghezza**	»	**1360**
Carreggiata posteriore	»	**1200**	**Altezza fiancale da terra**	»	**680**
Peso autotelaio con cabina guida	Kg	**400**	**Lunghezza cassone**	»	**2000**
Peso completo con cassone	»	**460**	**Larghezza cassone**	»	**1360**
Portata utile: . 2 persone +	»	**500**	**Consumo norme CUNA** lt. **5** per **100** Km		

DIREZIONE VENDITE Iso S. p. A. MILANO

CORSO PORTA NUOVA N. 16 • TELEFONI 665.850 - 665.890

Amongst the desirable consumer goods advertised the little Isetta speaks for itself. It has the sales appeal of small size

LITTLE BODY, BIG HEART

Convenience, Economy and More Room Than You Would Think in the Isetta

FORTUNATELY, personal tastes in most matters vary very considerably, and in Great Britain they vary over few things as much as cars. Some people might hesitate to admit that the B.M.W. Isetta is a car, but I for one would class it as such without doubt, and would go farther and add that it is a thoroughly agreeable little four-wheeler with comfortable width for two abreast, enough room for two adults and a child and, at a squeeze, the capacity to take three grown-ups for short distances. It must be admitted, all the same, that there is a distinctly aeronautical flavour about the construction and appearance and a motor cycle sound from the engine.

Being a novelty, as yet little known over here, it is a case of seeing and trying out to believe. Having covered perhaps 50 miles of main road in company with all the other traffic, to alight and glance back at the Isetta Motocoupe is to receive something of a shock. Could this diminutive bubble really have offered so much comfort and room, have handled like a car of twice its size and held its own, often at over 50 m.p.h.?

Convenience is high on its list of attributes. The big front entrance allows a passenger to walk in carrying a shopping bag or tennis kit with greater ease than he or she could enter an ordinary car door. The driver can climb out without disturbing the passenger and either one can step safely out on to the pavement, having parked in less than half the length of an ordinary small car, because the technique with the Isetta is to drive up to the kerb nose-first. A small but significant point concerned with the front entry is that you turn round and sit down normally without sliding along the seat, and thus your clothes remain comfortable without the need for tugging and shuffling.

The *de luxe,* smoke-blue Isetta used for a weekend recently was admired everywhere it went. It was equipped with screen wiper, sunshine roof, winking traffic signals, heater and radio. The four-speed (with reverse) gate-change gear box was easy to handle, and the steering and brakes were such as to give confidence at any speed of which the car is capable.

The lock is not as good as one might suppose, the

Park it at the kerb—end on—open the door without difficulty and step out, or in

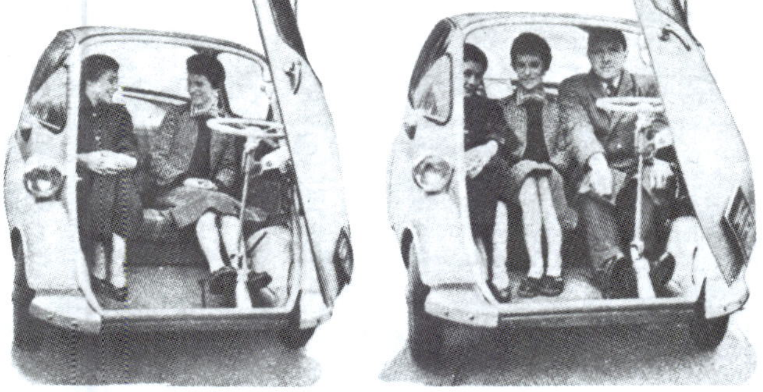

Room for two small ones with space to spare. Room for one big one and two small ones at a pinch

turning circle being 24ft. But this is quite sufficient for ordinary use. It came as no surprise that the back end was inclined to slide about on snow, but it followed the widely spaced front wheels safely enough. The very short wheelbase occasionally caused a rapid pitching motion over some kinds of rutted road surfaces, but this brought no more than a momentary upset to an otherwise dignified gait.

On the coldest mornings the engine started easily and as soon as the chill was off it idled well. Once the technique for a lukewarm start had been learned, there was no more trouble with under- or over-choking. For a single-cylinder four-stroke engine, the running is surprisingly smooth, and several passengers took it for that of a twin.

Photographs give a better impression of size than does the quotation of figures but, for those interested, the wheelbase is 4ft 10in, the track front and rear 3ft 11in and 1ft 8½in, the overall length 7ft 6in, and the height 4ft 4½in. The empty equipped weight is about 800 lb, and the ground clearance 6in.

If the standard Isetta could be bought here at its national price there would no doubt be queues of purchasers at Isleworth. As it is, this runabout is slowly

Where the carriageway takes two the Isetta can mostly make it three

finding favour as it becomes better known and its unusual but attractive bubble shape is seen about—this in spite of the import duty it carries in addition to £147 purchase tax.

Although there does not seem to be any need for more power or pick-up with the standard model, a little extra acceleration never seems to be amiss. Provided there is no loss of economy, an alternative 300 c.c. version of the Isetta, which has been mentioned in Germany for future production, may have added appeal for some owners. If I were thinking of detail improvement I should try to lower the shelf behind the seats, so as to give more stowage space without obscuring the rear view. Incidentally, the spare wheel is housed behind the driver's seat.

Seldom was there a handier little machine for its 245 c.c. and 55 plus m.p.g. No vehicle could have been more willing and convenient for the weekend shopping, football match, visit to the club and to friends, and it took me to and from the office in fine style and comfort.

M. A. S.

The B.M.W.-Isetta Coupe

WHILE visiting A.F.N. Ltd. at Isleworth recently I could not help being conscious of a " scuttling-about " by bubble-like vehicles, even though I was admiring Sebring and Le Mans Frazer-Nashes and Porsche Carrera coupés. These bubble-shaped objects were the little Isetta two-seaters of Italian Iso design and built under licence by B.M.W. in Munich, A.F.N. Ltd. being the importers. When the opportunity was offered to try one of these economy cars I readily accepted, for they are distinctly unorthodox in design and have the appeal of novelty.

Powered by a 245-c.c. pushrod o.h.v. B.M.W. single-cylinder engine, coupled to a delightful four-speed and reverse gearbox and mounted beside the rear wheels, the manner of going was anything but a novelty, it proved to be a revelation. With i.f.s. by coil springs and ¼-elliptic rear springs the ride is quite good for such a light vehicle and the very light steering has a pleasant understeer characteristic. One of the pleasures in driving the Isetta was its control and handling being quite orthodox. So many economy cars call for manipulation of some comic feature, either a special type of gear-change, an unusual method of starting, very odd steering or a lack of brakes. The Isetta suffers from none of these and everything is worked by normal car practice, even the starter being brought into play by merely turning the ignition key beyond the normal " on " position, as with most modern German cars, while headlights of surprising power are dipped by a lever on the left of the steering column, a similar one on the right operating the direction indicators. Seating two people with comfort and room for three at a squeeze, with a vast parcel shelf behind the seats, this little 245-c.c. runabout is ideal for shopping trips or inter-village communication out in the country. Providing the engine is taken up to its maximum of 5,800 r.p.m., and the gearbox is used to the full, one is never in the way of other traffic as with many economy cars, in fact a 45 m.p.h. cruising speed, with another 5 m.p.h. in hand, deals with the average British family saloon and " Mimsers " get in the way.

In city traffic the Isetta is a joy, for with its very crab-track layout once the front is through a gap there is no need to worry about the rear, though at times one gets a feeling of being in the dark when squeezing between two double-decker buses, and, of course, with the way the average driver creeps along there is always room for the Isetta to pass by on the inside in busy streets. The parking problem is immediately halved when using an Isetta for it is only a few inches over 7 ft. in overall length and if the nose is allowed to overhang the pavement a little it can be driven into a parking gap at right angles to the road, needing only a 4 ft. 6 in. gap for this manoeuvre. Having a single door in the form of the complete front of the car, hinged on the left, you then step forward out on to the pavement, much to the consternation of passers-by. As the whole of the upper half is of glass and Perspex, visibility is literally " all round," which makes the occupants rather conspicuous, but the " bubble " shape of the Isetta attracts attention anyway, so it is no added hardship.

The Isetta must surely represent the most successful way of providing transport at low cost while avoiding the complications of a motor car and the disadvantages of a motor-cycle or scooter, while its cornering power and general safety in handling, relative to its size and power, classes it with cars of the highest order.

When I heard that the Italian versions of these little vehicles had averaged 50 m.p.h. from Brescia to Rome in the Mille Miglia, I found it hard to believe. Now I have sampled the B.M.W. version I can well understand how it was done.

Having covered 350 miles in the course of a quiet weekend I suddenly remembered it was supposed to be an economy car so the tank was drained and 62 miles were covered on a gallon, cruising at 30 m.p.h. uphill and 40 m.p.h. downhill, around North Hampshire. Economical and lively, as well as being enormous fun to drive, though a bit noisy by some standards, was the final verdict, and fantastically safe. If the Government would remove Import Duties and other money-grabbing rackets there would beone in every home, but even so anyone who delights in something that is not only different but also functional adn practical, should call in at Isleworth.

D. S. J.

JOHN BOLSTER TESTS THE

BMW Isetta

The Isetta is a most intriguing little car. Originally built in Milan, it is now also constructed under licence by Velam in France and BMW in Germany. It is the latter version that I have recently been using, and it differs from the other two in several respects, particularly in having a four-stroke engine.

The whole design has been dictated by practical considerations, notably ease of parking. Thus, the car is so short that its overall length is little more than the width of a Cadillac. It can, therefore, be parked at right angles to the pavement, whereupon the whole front panel, including the screen, opens to permit easy entry and exit. The steering wheel swings out of the way with the door, having a universally-jointed column.

The basis of the Isetta is a box and tubular section frame, to which the tubular body structure is directly secured for extra strength. In front, the steering gearbox is attached to the left corner, and is coupled to the stub axle just behind it by a short link. A plain track rod commands the offside stub axle. This ultra-simple layout is feasible because the independent front suspension is on the Dubonnet system. The suspension units, containing helical springs and hydraulic dampers, turn with the steering and carry the wheels on swinging arms, unsprung weight being thus at a minimum.

COURSE of instruction before taking over the Isetta Motocoupé. The unique entrance and compact dimensions are well shown.

Flat and unencumbered, the floor leads back to a comfortable bench-type seat. The engine lives behind this on the right side of the car, and is in unit with a four-speed gearbox. The drive then goes through a very short universally jointed shaft, the engine, box and shaft lying "across" the car, so to speak. The power is transmitted to the back axle by a duplex chain, which is enclosed in a light alloy housing. The axle itself is solid, without a differential, which is rendered superfluous by the narrow rear track of only 1 ft. 8½ ins. This axle is sprung on a pair of quarter-elliptic springs, and attached to vertical telescopic dampers.

The engine is a single-cylinder unit, cooled by a blower and ducting, and at first sight is of motor-cycle appearance. In fact, however, it is of typical car construction, with an external flywheel, counterbalanced crankshaft, and normal wet sump pressure lubrication. The dynamo-starter is in unit with the fan, and spins the engine rapidly without the aid of a compression release.

On taking one's seat, the driving position at once appeals. One somehow does not notice the absence of a bonnet, and heavy traffic is not at all frightening. Only the front small windows pivot, the remaining large, curved, transparent area being fixed. In bright sunshine, it tends to become a bit cosy, but a flick of the hand has the sunshine roof folded back.

The clutch is rather on the fierce side, and needs practice. The gear lever, on the side of the body, allows instantaneous changes up or down on full throttle, but is slightly stiff when used in a leisurely manner. The gear ratios are very well chosen, and, of course, so small an engine requires the intelligent use of the lever, which somewhat confusingly has all its positions "back to front".

Perhaps the actual performance needs putting in the right perspective. If one simply takes the acceleration graph and compares it with that of any full-sized new car, it certainly makes the Isetta seem sluggish. Yet, on a typical journey, a large percentage of the vehicles encountered are elderly, and the Isetta can swish past the average pre-war "seven", "eight", or "ten". For instance, I had a run down to the seaside in it, with two children and a good deal of impedimenta. I cruised easily at 40 to 50 m.p.h., and passed more cars than passed me. It is true that every real hill needed a change of gear, but who begrudges this to a 245 c.c. engine?

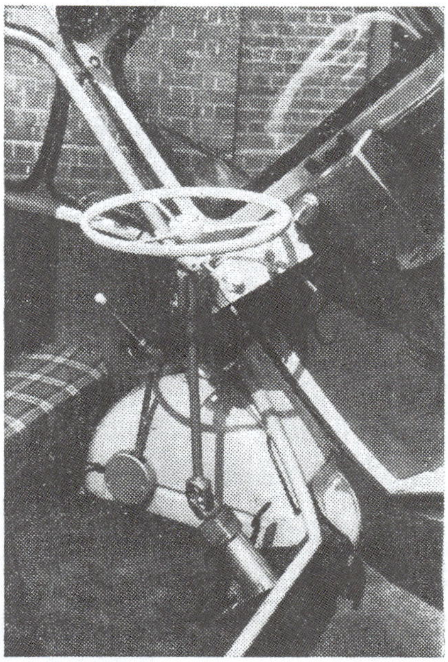

INGENIOUS in the extreme, the steering wheel is mounted on the door, and opens with it. The gear lever and hand-brake are on the left.

Quite the most remarkable feature is the comparative silence Most air-cooled cars are noisy, but this little machine is quite reasonable in this respect. I have indeed, driven a great many water-cooled cars that made more row. If one keeps the revs. down a little, the radio can be enjoyed without disturbance.

The steering of the Isetta is superb. Very light, it is completely accurate, and there is no conscious effort in keeping straight, even at maximum speed. There is a slight understeering tendency, which ensures stability, and it is literally true that one can completely ignore the shortness of the car. It would, however, be a miracle if really comfortable suspension could be achieved with a wheelbase of only 4 ft. 10 ins. The ride is, in fact, more than praiseworthy for so tiny a vehicle, but some types of rippled surface can produce a noticeable pitching sensation, though this never becomes unpleasant.

The brakes are very large, having regard to the light weight and fairly moderate speed. As would be expected, they have complete mastery of any situation. These are properly engineered, with full hydraulic operation, and no adjustment was required during a strenuous testing period. I stress this point, for reasons that will be abundantly clear to those who have experienced certain baby-car brakes, of the "string and wire" variety! At low speeds, some "winding up" of the front suspension can be felt under abrupt braking.

Unlike some small cars, the Isetta seems to be running well within itself, even at a steady 50 m.p.h. cruising speed. The engine characteristics are such that it does its best work in the medium and upper ranges, but the power curve then falls off very rapidly, and the slightest over-revving slows the acceleration markedly; albeit the quoted maxima on the gears can be considerably exceeded without mechanical danger.

BMW Isetta—*continued*

Anybody who buys an Isetta must be prepared to be the centre of attraction wherever he stops. I am lucky enough to drive many of the world's glamour-cars, but this little machine collected a bigger crowd than any of them. Even the police stopped me, with many apologies, because they simply had to have a better look! In my family circle, the tiny vehicle was immediately adopted with affection, and my schoolboy son christened it "Egbert", a name which it retained throughout my "ownership".

ACCESS to the 250 c.c. o.h.v. engine is afforded by removal of a single body panel on the offside.

I do not think that the Isetta competes with our low-priced three-wheel minicars. Having four wheels, it pays the full car tax, and its price rather puts it out of their class. I think it will sell largely to people who already have a Mercedes-Benz, a Jaguar, or a Bentley in the garage. For them, the three-wheelers are too crude, and the typical small family car lacks distinction. The Isetta is just as well built as any big car—the name BMW guarantees that. Its finish and appearance make it a second car that will appeal to the man or woman with an educated motoring palate.

SPECIFICATION AND PERFORMANCE DATA

Car Tested: BMW Isetta Motocoupé, price £292 (£439 7*s.* 0*d.*, including P.T.). Wheel discs and radio extra.

Engine: Single air-cooled cylinder 68 mm. x 68 mm. (245 c.c.). 12 b.h.p. at 5,800 r.p.m. 6.8 to 1 compression ratio. Pushrod operated inclined overhead valves. Motor-cycle-type carburetter, coil ignition.

Transmission: Single dry-plate clutch. Four-speed constant-mesh gearbox in unit with engine, with remote control on left side of body, ratios 6.24, 8.18, 11.94, and 23.22 to 1. Final drive by enclosed chain to solid rear axle.

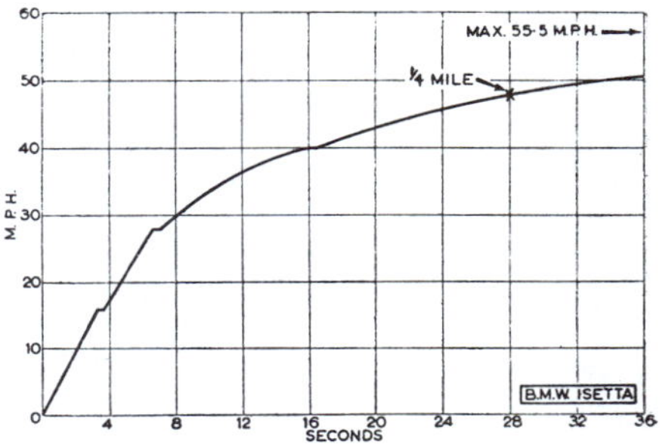

ACCELERATION GRAPH

Chassis: Box section and tubular frame with superimposed tubular body frame and stressed panels. Dubonnet-type independent front suspension incorporating helical springs and hydraulic dampers. Rear suspension by quarter-elliptic springs and telescopic dampers. Bolt-on split rim disc wheels, fitted 4.80-10 ins. tyres. Hydraulic brakes in 7 ins. drums.

Equipment: 12-volt lighting and starting. Speedometer. Flashing direction indicators. Heater. Reserve fuel tap. Electric wiper. Radio (extra).

Dimensions Wheelbase 4 ft. 10 ins.; track, front 3 ft. 11¼ ins., rear 1 ft. 8½ ins.; overall length, 7 ft. 6 ins.; width, 4 ft. 6¼ ins. Turning circle, 29 ft. Weight, 6½ cwt.

Performance: Maximum speed, 55.5 m.p.h. Speeds in gears, 3rd 40 m.p.h., 2nd 28 m.p.h., 1st 16 m.p.h. Standing quarter-mile, 28 secs. Acceleration, 0-30 m.p.h. 8.2 secs., 0-40 m.p.h. 16.2 secs., 0-50 m.p.h. 34.6 secs.

Fuel Consumption: 62 m.p.g.

AN IMPROVED BMW ISETTA

CHANGES TO BODY SHELL AND SUSPENSION : BRIEF ROAD IMPRESSIONS

A DE LUXE version of the BMW Isetta, introduced at the Frankfurt Motor Cycle Show in October of last year, and already produced in considerable numbers, is changed in several important respects from its predecessor. In appearance it has been partially "un-bubbled" and its windows brought closer in line with car practice. Thus, it now has conventional rear and side windows, the latter having large area sliding panels to provide better ventilation than the standard model's small, hinged quarter lights.

Externally the improved Isetta is distinguished by revised side and rear windows. The dual colour scheme is optional. A new front suspension layout and more supple rear springs are introduced

Baggage space on the deck behind the seat has been increased by lowering the deck level slightly. On export Isettas two fresh air inlets are added to the door panel. A dual colour scheme is now available, and the colours are divided by a bright metal strip around the waistline.

New body style offered on the improved chassis is the cabriolet, which has a folding hood covering the rear deck to supplement its folding sun-roof.

Modifications to front and rear suspensions have markedly improved the ride. Longer quarter elliptic springs for the rear axle have been accommodated by abbreviating the chassis frame, the wheelbase remaining as before. At the front, the wheels are still carried on fore and aft swinging arms, rear-hinged, but long, vertical coil springs enclosing double-acting hydraulic dampers replace the previous Dubonnet arrangement, and vertical wheel travel is increased by 32 per cent.

Even more important than the softer ride now provided is the elimination of a pronounced pitching motion which could be induced sometimes in the previous model when braked hard on a bumpy surface.

Inside the new Isetta one has more privacy from the reduced depth of the rear window, although the outward aspect is perhaps not so attractive from that angle. The front pillars are still thick, but driver visibility through the side windows is much improved.

The revision of the suspension is at once apparent on the road, and the little vehicle negotiated pot-holed side roads quite fast with surprising stability. Leg room for the driver is still insufficient for real comfort on a long run, and an increase of even two or three inches would be appreciated.

Although the example tried was fitted with the larger of the alternative power units (295 c.c. as compared with 245 c.c.), it had covered only a very small mileage and was too new and stiff to extend. It was impossible, therefore, to match its performance in terms of maximum speed and acceleration against that of the 245 c.c. model, of which a Road Test appeared in *The Autocar* dated November 4, 1955. Undoubtedly it is livelier.

Isetta KLEINWAGEN

New, efficient and safe 4-wheel automobile. Built in Germany by renowned Bayerische Motoren Werke A.G., Munich. For generations builders of luxury cars, motorcycles and aircraft engines.

CONVERTIBLE

despite its hardtop (center rolltop).

TWO CARS PARK IN THE SPACE OF ONE

Isetta Kleinwagen has its ONE door in front. The frontal access and the compactness of the car will make parking easier and allow frontal parking in many localities.

Two adults and a child ride comfortably on its upholstered seat. Ample baggage compartment.

ECONOMY and EFFICIENCY

Isetta Kleinwagen travels 60 miles on a gallon of gasoline, climbs 30 degree roads in first gear, allows a maximum speed of 53 miles per hour.

No Antifreeze

Isetta Kleinwagen has one cylinder, 4 cycle motor with forced air-cooling system, 13 HP, Rear axle drive, 4 forward shifts and one reverse gear.

Priced at approx. $1,000 fob port of entry, Isetta Kleinwagen will soon be available through dealers.

Sole Importers for U.S.:

FADEX COMMERCIAL CORPORATION

136 LIBERTY STREET

NEW YORK 6, N. Y.

SPECIFICATIONS FOR BMW ISETTA KLEINWAGEN

Motor	BMW, 1 cylinder, 4 cycle, with forced air cooling system, overhead valves
Bore	72 mm
Stroke	73 mm
Piston Displacement	300 ccm
Compression Ratio	7:1
Power	13 HP -- RMP 5200
Lubricating System	Gear pump, forced feed lubricating system
Clutch	Single disc, dry clutch
Motor Location	On the right side of car in the rear, behind seats
Starter	12 V, 130-W, with voltage regulator
Transmission	BMW 4 forward and 1 reverse, gearbox attached to motor
Rear Axle Drive	Resilient mounted driveshaft and a totally enclosed, fully adjustable chain drive in oil bath
Chassie-Frame	Strong tubular steel construction
Axles and Suspension System	Front wheels individually sprung by use of coil springs and shock absorbers. Rear wheels mounted on conventional axle, close wheel space, 2 quarter-elliptical leaf springs and telescopic hydraulic shock absorbers
Wheels	2 piece steel discwheels, to facilitate tire changes.
Tires	4 tires, plus 1 spare, in super-comfort size 4.80 x 10"
Steering	Steering wheel, reduction steering gear, smallest turning circle approx. 24 ft.
Brakes	4 wheel hydraulic brakes, brake drum diameter 7", total braking area 49.9 square inches
Dimensions	Width of front wheel track 47.25" Width of rear wheel track 20.5 " Wheel base 58" Over-all length 7½ ft. Over-all width 4½ ft. Over-all height 52.5"
Weight	Empty 770 lbs. Carrying capacity 500 lbs.
Speed	Maximum speed 53 MPH
Gasoline Consumption	About 60 miles per gallon
Gradeability	In first gear over 30%

BMW BY

(Left)—The steering column of the Isetta "breaks" and swivels out when the large, single door is opened. (Technical description of the car appeared last September.)

SINCE the last issue went on the counters we have undergone a most salutary experience. Stung perhaps, by O'Hagan's cutting remarks about Dwarf cars last November, the Irish Distributors for BMW hinted that we might like to take over the reins of their new Isetta which had just arrived at Cork docks. In accepting the offer we privately decided to really give this Dwarf "the works", and the week that followed saw Ireland's first BMW Isetta facing a period of testing such as we have visited upon few cars, be they big or small.

The most interesting effect of the experiment was not what we did to the Isetta, but what it managed to do to us. It taught us that you can fully enjoy long drives in a pint-sized vehicle without the slightest feeling of embarrassment or discomfort. In addition we were provided with just the right sort of background for further discussion (as promised last month) on the vexed question as to whether or not the Dwarf car has a future stake in the Irish market.

The Isetta's ace-in-the-hole is the fact that it is made and backed by one of the oldest and most revered motor firms in the world, Bayerische Motoren Werke of Munich who once made a series production car and guaranteed it against major defects for 100,000 miles. It is hardly surprising that the BMW is still regarded as the German Bentley.

The Isetta, it will be recalled, started life as a brilliant piece of design engineering by a couple of Romans who had the vision, but not the backing to get into production. When BMW's scout saw how the first hand-built prototypes waltzed through their classes in successive Mille Miglias he reported back accordingly. A bid was made, and ended in BMW's securing the manufacturing and marketing rights for all Western Europe outside Italy. The intelligence of the Germans was not so much that they recognised a trail-blazer, but that they had the good sense to take over the original Isetta without trying to imitate it.

Most Dwarfs are scaled-down adult cars whereas the Isettà is something radically different . . . with the professional touch very much in evidence. It is perfectly finished, carries the right sort of accessories, and is as strong as a pocket battleship.

The tiny 300 c.c. air-cooled single cylinder engine developes a maximum of 13 b.h.p. at nearly 6,000 r.p.m. and is ably assisted by as pretty a gear-box as we have ever tried. Performance of the acceleration-against-the-clock variety is relatively unimportant in such marginal transport. The thing that matters is the ability to 'keep up' with conventional type traffic under fast touring conditions on the straightaway as well as on the hills, and to keep ahead of the others at the green lights.

You have to measure up these essentials while under maximum load because the motocoupé spends most of its travelling life fully laden. The Isetta achieves what we can only describe as near-miracles in measuring upto the standards set, we proved that it can cruise indefinitely with most of the others in the 8 h.p. class and scamper up inclines as quickly as cars with three times its engine capacity.

The only stipulation required to gain such results is that intelligent use of the gear-box must be combined with determination to keep up the engine revs. when competitive motoring is called for. At the same time we cannot over-emphasise the incredible degree of slow-speed refinement which the BMW provides for those who prefer it.

It is difficult to reconcile accomplishments of this order (on a car of this size), with the sort of durability today's small car owner has been educated to regard as normal. Needless to say a week's testing doesn't guarantee any answers, but we thought a lot about the point. Chassis and bodywork will, we feel, go the full limit with any adult low-priced car with ease, and BMW's name on the engine and transmission is backed up by the usual six months warranty (or 7,000 miles, whichever comes first). More practically speaking, the engine has a big fat "pot," only the Volkswagen has a bigger in the upto £575 class, and the bore to stroke ratio is dead 'square' so that related piston speed is low. On paper the unit doesn't work any harder than a piston assembly on a four cylinder engine, it's cleverly geared down, and is correctly lubricated by a pressure pump drawing from an orthodox sump.

This brings us to the matter of comfort. A highly important one too, for if there is anything the public will not wear, it is discomfort. The Isetta, having to concern itself with the legitimate needs of two adults only, can afford to give excellent leg and head room, plus a width at elbow height (46½ inches) which is slightly more than in the Volkswagen! In the riding sphere it would be preposterous to claim that a car with such a limited wheelbase could ride as well as one with 50 per cent more distance between front and rear axles. The wheels meet the same corrugations but the "recovery rate" is so hastened as to give a more pronounced reaction in the form of a mild pitch. Only a deceiver would say that you don't notice this, yet the truth is that the BMW is now so softly sprung at the rear that the characteristic does not irritate in the slightest. So far as we were concerned the ride was much, much better than we had been expecting.

It's just another case of not applying copybook methods of comparison . . . you must judge the Isetta for what it is. Nowhere does this arise more noticeably than in the noise section. You couldn't expect to have as low a noise factor in a one-piece bubble (with the engine churning within inches of your tail) as in a complex weighty saloon with an insulated bulkhead and two or three feet of miscellaneous equipment between the passengers and the 'works'.

Once again the Isetta springs a surprise, its noise level is not high enough, or harsh enough, to intrude on a normally pitched tête-à-tête at 45 m.p.h. The uncanny thing about the whole business is that when you emerge from the car · it suddenly looks much smaller, and sounds far noisier than you thought when you were actually travelling in it!

The BMW Isetta is certainly one Dwarf in point which has no need to cover up any shortcomings . . . as O'Hagan suggested . . . with gadgetry like opening roof, cigarette lighter and screen washers, though we could not depart from it without recording that the great thickness of the screen pillars almost cancels

ADOPTION

(Right)—From this angle the BMW Isetta looks odd. The view was chosen to show the rear wheel arrangement by which the vehicle qualifies as an 8 h.p. . . . and consequently gets 450 miles worth of petrol ration each month. The luggage rack is an extra.

out the manœuvrability aspects, or that we felt anything but distaste for the effort required to clamp home the door (necessary to align the steering column) and for the so-called heater . . . did we say heater? Incidentally the makers convey an impression that you can adopt a new method of parking with the Isetta, i.e., by driving straight into the kerb. Unfortunately they ignore the invitation to side panel damage when the 'other fellow' starts backing up for space to move out, as bumpers are fitted to front and rear only.

Where does all this leave us with the Dwarfs? What are our feelings now about their chances on the Irish market?

Before the Isetta incident we were definitely sceptical. After it we are a lot less sceptical . . . but not entirely convinced that its future here is only a production problem. Let's quote the odds as we see them.

In the first place it is worthless to try to apply Continental experiences, as the Dwarfs will all be around £100 dearer in Ireland that at home. You can thank levies, duty, and freight charges for part of this, and the high overheads of native assembly for the rest. The Dwarf costs pretty nearly as much, per unit, to assemble as the adult car.

The various distributors are not blind to the importance of initial price and we have no doubt that in one case at any rate, the Irish quoted figure is barely profitable. The battle-cry of the Dwarfs is one of economy. In initial price, in upkeep, in depreciation, and in servicing. When it comes to petrol consumption they speak a language of their own. The BMW for instance, gives an easy 65/75 miles per gallon (pure; for oil goes into sump separately) and no Dwarf worthy of its salt would be caught guzzling more than a gallon to 55 miles. On the tax side they do not fare so well. The three wheelers like the Heinkel, get by on £8 annually. The four wheelers get it in the neck with the full 8 h.p. (£13) rate. One assumes that the brand names of the various Assemblers will be enough to guarantee an equitable and adequate spares and service background for second and third owners when their turn comes, and this in turn will have a direct bearing on depreciation as influenced by re-sale value.

Examining the position of the Dwarfs in the hard, unadulterated light of what they offer for the money you would be inclined to vote on their success without much hesitation. Unfortunately, another obstacle lies in their path which could hobble them without any reference to their qualities and attractions.

The present state of the car market is reflected in large stocks of secondhand cars, mostly marked down at extremely keen prices. The garage-men will tell you that they could sell any number of new cars if they could take in more trade-ins. . . . Will not the Dwarfs come into direct competition with the really excellent used-car values now available? And will not the big Assemblers endeavour to prevent their Dealers from getting tied in with something that may increase their liabilities in the secondhand line, assuming as has been shown, that new car sales are largely dependent on used car clearances?

If we were the Agents for such a car as the BMW we would concentrate our every effort on pushing through a hire-purchase scheme which would rival the model itself in attractiveness. And, that, judging by the number of people who clustered around the Isetta in every town we visited, would be no easy task!

There is of course the cherished hope among most of the Dwarf distributors that their vehicles will have a particular appeal to businessmen (as 'seconds' for wives) and to present scooter owners, looking for something weatherproof. Continental makers like DKW and Fiat reckon that the latter market will be big enough to justify their entry into the ranks of Dwarf makers (and not, repeat NOT, as replacements for existing products!). But soliloquising on these lines is strictly off-beat in a country labouring under a diminishing, as against an increasing, economy. Ambitious scooterists here feel that they will be lucky to maintain the *status quo,* and jolly few dare to think any further ahead than that.

Of course we could be wrong for visualising difficulties of the type mentioned. We would have been the last to believe that there is a good secondhand market for expensive used quality cars in Ireland, costing upwards of £800. There isn't? Ask any Rover dealer! ■

FOOTNOTE: The expression "Dwarf" is applied to those vehicles which come halfway between the scooter and the orthodox small car, and it is not intended to be any more derogatory than the term "Baby-Car" which is popularly used in references to models in the £400 to £500 range.

Buyer's Guide

Price	£340 (approx.)	Basic Ration (Month)	6 galls.	Weight Distribution	54p.c.F. 46 p.c.R.
,, with Heater	£340 ,,	Lighting System	12 volt	Windscreen Area	465 sq. ins.
Annual Tax	£13	Brake Lining Area	49 sq. ins.	Starting Handle	Yes
Annual Insurance (Comphse.)	£22-17-0	M.P.H. at 1.000 r.p.m. (Top)	21	Paintwork	Baked Enamel
Cubic Capacity	295 c.c.	Overall Height	52.6 ins.	Steering (Lock to Lock)	2¼ turns
Bore	72 m.m.	Overall Length	90 ins.	Servicing Interval	1,000 miles
Stroke	73 m.m.	Overall Width	54.3 ins.	Parcel Shelf	Yes
Compression Ratio	7 to 1	Wheel Base	59 ins.	Interior Light	No
Brake h.p.	13 @ 5,800	Track (Minimum)	47 ins.	Exterior Boot	No
Max. Torque	26 ft. lbs.	Hip Room (Front)	47 ins.	Bumper Over-riders	No
0—30 m.p.h.	10.5 secs.	Hip Room (Rear)	—	Warranty Period	6 months
0—50 m.p.h.	35 secs.	Head Room (Rear)	—	Replacement Engine Scheme	Yes
Max. Top Gear	53 m.p.h.	Tyre Size	4.80 x 10	Independent Suspension	Yes (front only)
Max Inter. (or 3rd)	40.3 m.p.h.	Turning Circle	26 ft.	Service Voucher Scheme	Yes
Target m.p.g.	76 m.p.g.	Kerb Weight	5½ cwts.		

The Motor Road Test No. 12/57

Make: B.M.W. **Type:** Isetta 300

Makers: Dunsfold Tools, Ltd., Brighton, Sussex, and Bayerische Motoren Werke, A.G., Munich

Test Data

World copyright reserved: no unauthorized reproduction in whole or in part.

CONDITIONS: *Weather:* Cool and showery with light wind. (Temperature 42°-46° F., Barometer 29.8-29.9 in. Hg.). *Surface:* Damp tarred road. *Fuel:* Standard-grade pump fuel (approx. 80 Research Method Octane Rating).

INSTRUMENTS

Speedometer at 30 m.p.h.	15% fast
Distance recorder	3% fast

WEIGHT

Kerb weight (unladen, but with oil, coolant and fuel for approx. 50 miles)	7⅝ cwt.
Front/rear distribution of kerb weight	55/45
Weight laden as tested	10⅞ cwt.

MAXIMUM SPEEDS

Flying Quarter Mile

Mean of four opposite runs ...	50.1 m.p.h.
Best one-way time equals	51.1 m.p.h.

"Maximile" Speed. (Timed quarter mile after one mile accelerating from rest)

Mean of four opposite runs ...	49.7 m.p.h.
Best one-way time equals	50.8 m.p.h

Speed in gears

Max. speed in 3rd gear	40 m.p.h.
Max. speed in 2nd gear	28 m.p.h.

FUEL CONSUMPTION

84 m.p.g. at constant 20 m.p.h. on level.
78½ m.p.g. at constant 30 m.p.h. on level.
64 m.p.g. at constant 40 m.p.h. on level.

Overall Fuel Consumption for 828 miles, 14¼ gallons, equals 58.1 m.p.g. (4.86 litres/100 km.).

Touring Fuel Consumption (m.p.g. at steady speed midway between 30 m.p.h. and maximum, less 5% allowance for acceleration) 60.8 m.p.g.

Fuel Tank Capacity (maker's figure) 2.8 gallons

STEERING

Turning circle between kerbs;

Left	30 feet
Right	30 feet
Turns of steering wheel from lock to lock	2½

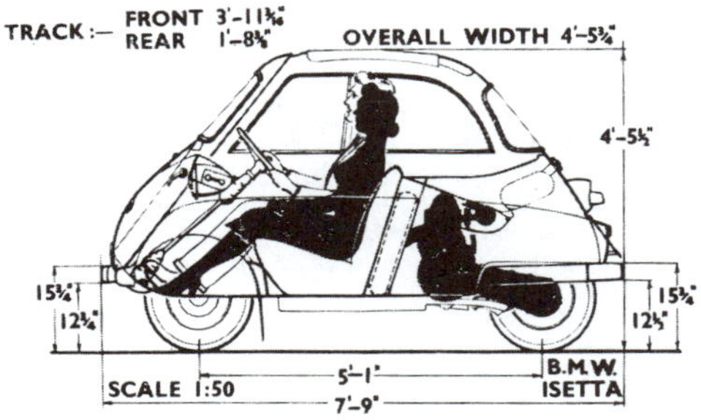

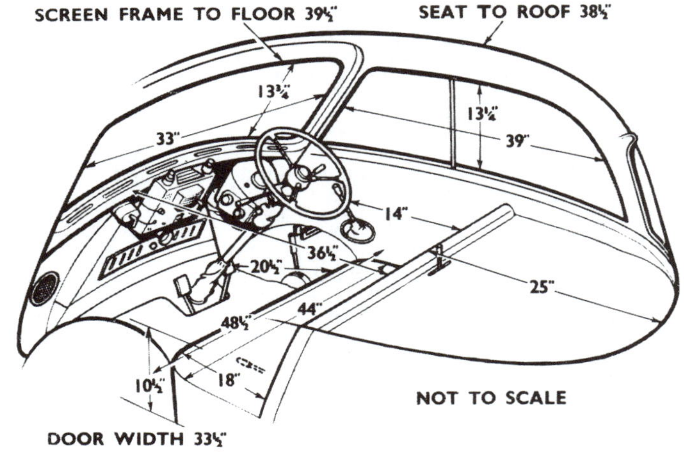

ACCELERATION TIMES from standstill

0-30 m.p.h.	12.0 sec.
0-40 m.p.h.	23.5 sec.
Standing quarter mile	29.6 sec.

ACCELERATION TIMES on Upper Ratios

	Top gear	3rd gear
10-30 m.p.h.	24.2 sec.	15.0 sec.
20-40 m.p.h.	30.7 sec.	—

BRAKES from 30 m.p.h.

0.87g retardation (equivalent to 34½ ft. stopping distance) with 100 lb. pedal pressure
0.64g retardation (equivalent to 47 ft. stopping distance) with 75 lb. pedal pressure
0.35g retardation (equivalent to 85 ft. stopping distance) with 50 lb. pedal pressure
0.16g retardation (equivalent to 188 ft. stopping distance) with 25 lb. pedal pressure

HILL CLIMBING at sustained steady speeds.

Max. gradient on top gear	1 in 20.3 (Tapley 110 lb./ton)
Max. gradient on 3rd gear	1 in 13.5 (Tapley 165 lb./ton)
Max. gradient on 2nd gear	1 in 8.9 (Tapley 250 lb./ton)

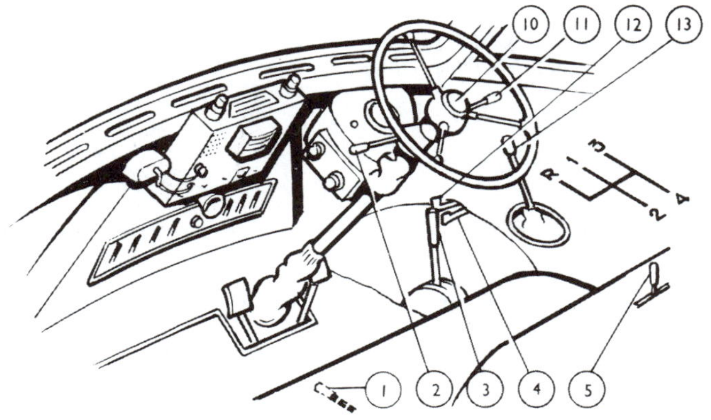

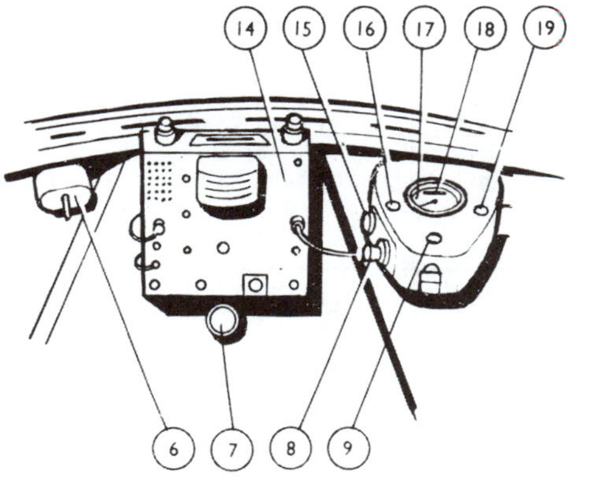

1, Heater control. 2, Headlamp dip switch. 3, Handbrake. 4, Ventilator control. 5, Petrol tap. 6, Windscreen wiper. 7, Cigar lighter. 8, Lights switch. 9, Direction indicators warning light. 10, Horn button. 11, Direction indicator switch. 12, Choke control. 13, Gear lever. 14, Radio. 15, Ignition and starter switch. 16, Dynamo charge warning light. 17, Speedometer. 18, Distance recorder. 19, Headlamp main beam indicator light.

— The B.M.W. Isetta 300

Real compactness makes the Isetta ideal for town use, its length only a little greater than the width of large cars. Front ventilator, wing mirror and luggage grid shown here are non-standard features.

A Practical Miniature of Extreme Economy and Lively Performance

In Brief

Price £265 plus purchase tax £133 17s. 0d. equals £398 17s. 0d.

Capacity	298 c.c.
Unladen kerb weight ...	7⅝ cwt.
Acceleration:	
20-40 m.p.h. in top gear ...	30.7 sec.
0-40 m.p.h. through gears	23.5 sec.
Maximum top gear gradient	1 in 20.3
Maximum speed	50.1 m.p.h.
Maximile speed	49.7 m.p.h.
Touring fuel consumption ...	60 m.p.g.

Gearing: 9.3 m.p.h. in top gear at 1,000 r.p.m.; 19.4 m.p.h. at 1,000 ft./min. piston speed.

IN the two and a half years since *The Motor* published the first road test report on the Isetta, much has happened to this unorthodox vehicle. In the first place a licence for the original Italian design with which our report was then concerned was acquired by the B.M.W. works of Munich, and the design modified principally in the matters of front suspension and power unit. The two-cylinder, common combustion chamber, two-stroke engine was exchanged for an adapted version of the B.M.W. single-cylinder four-stroke, of 245 c.c. capacity with ducted fan cooling. Many thousands of these machines were sold, the engine latterly being enlarged to 298 c.c. A change in body styling also occurred recently, with a less pronounced "bubble" at the back, sliding windows and more parcel space.

Finally, and most important, arrangements have been made for the manufacture of the B.M.W. Isetta in Britain, with right-hand drive. The subject of this report is a German-built prototype of the British Isetta which is revealed to the world at this week's Geneva Show.

One further changed circumstance might perhaps be included: The fact that, for Londoners at any rate, the front-opening door is no longer a novelty. Whilst this feature has its own advantages, particularly for the congested traffic of the city, it seems more likely that the present fuel situation has been mainly responsible rather in the face of such unorthodoxy, for the rapidly expanding sales of these vehicles.

Economy at Speed

With fuel scarce and expensive, a well-contrived small car must undoubtedly benefit given three conditions; that it shall fill the position of a larger car in reasonable comfort and in performance, and that it shall still be economical when driven to keep up with normal traffic. In the case of the Isetta these conditions appear to be amply fulfilled. The extent to which the model on test was used, and hard-driven, can be judged by the results of fuel consumption measurements. At a steady 30 m.p.h., which under emergency conditions many people are prepared to accept as a cruising speed for all but long journeys, a figure of comfortably more than 80 m.p.g. can be achieved. An accelerator with a definite "second pressure" makes this easier to realize in practice. In the course of driving by several members of our staff which varied for the most part between energetic and very hard, checks upon overall consumption showed a worst figure of 54.5 m.p.g. and a best of 62.4 m.p.g. This overall consistency seems to be a direct result of using a four-stroke engine of generous size. Avoiding the very small two-stroke's tendency to give disappointing economy under hard driving, the latest Isetta also shows up well in comparison with the 245 c.c. model with which we had some experience a year ago. Less dense, and slower-moving traffic may have a slightly favourable effect, but even in town conditions the 300 c.c. engine reduces the need for frequent downward gear-changes or full-throttle acceleration.

An added financial economy can be achieved by using commercial-grade petrol, which has only a small noticeable effect on clean opening-up at low speeds, and normal four-stroke lubrication imposes no extra charge for oil in the fuel supply.

When performance is needed, at least within the fairly leisurely limits of cruising speed now prevalent, acceleration times that on paper are equally leisurely tell rather less than the whole story. The fact that the great majority of drivers habitually use much less than the full performance of their cars is well known, moreover, and the job of keeping up with or overtaking other cars in the Isetta is made much easier by a four-speed, all-indirect gearbox so pleasant to handle that sheer laziness could be the only excuse for not using it.

Allowance must certainly be made for the rather unusual characteristics of a single-cylinder engine with a large fly-wheel, causing an appreciable time lag

A small change in the British-built version will be new headlights adopted with the Lucas electrical equipment.

The B.M.W. Isetta

Four wheels impose the car rate of tax, but make for excellent stability in all conditions. The latest model has more orthodox windows than its predecessor.

before the clutch can be engaged to move the car away from rest. The only fault of the gearbox appears also at this juncture, in occasional reluctance to find first gear. Final judgment on the mechanism, which at the time of testing had only just been converted from left-hand to right-hand drive, must wait until production examples are available. With the car under way, it is more or less impossible to crash the motorcycle-type gearbox, and the change is as fast as the lever can be moved. An improved and smoother clutch has been developed since the time of the first imports to Britain. So far as the controls are concerned, there is indeed little of the unorthodox miniature car in evidence. The luxury of key-starting is provided for a Siba Dynastart, which normally turns over even a cold engine rapidly and silently.

Long Week-Ends

In normal conditions, starting is easy from hot or cold; an air-cooled engine both warms up and cools down quickly, so that the choke is often needed, but only for a very short time. The gravity-feed fuel tank has a tap on the parcel shelf which it is desirable to turn off when the car is to be left for more than a few minutes, and (there being no fuel gauge) the opportunity has been taken to provide a reserve of $\frac{3}{8}$ gallon. With this included, the tank allows a range of about 200 miles.

For the majority of owners the occasions when a non-stop journey of that length arises are likely to be rare. Mechanically the Isetta appears more than robust enough for continuous hard driving, and one member of our staff covered a full 600 miles in a week-end, including a stretch of 140 miles with no more than a quick change of drivers. In all, the test car covered more than 1,000 miles, with no more trouble than an intermittently flooding carburetter. The human factor is really the one which would normally limit the car to local or suburban use, for an air-cooled single cylinder in close proximity to the rather confined interior inevitably produces more noise than the motorist is used to. The noise level is closely related to the speed and throttle opening. Gentle cruising, or main-road work with a following wind are comparatively restful, but a strong head wind has a pronounced effect on performance and the noise rises accordingly.

On the score of riding comfort little fault can be found, in spite of a wheelbase of only five feet. Recent detail work upon the suspension, including long rear springs, has enormously improved the ride by comparison with earlier models. The former high-frequency pitching tendency, magnified by the position of the occupants almost over the front wheels, has been eliminated. With four tracks making the job of dodging pot holes almost impossible, it is particularly useful that the Isetta takes really bad, almost "colonial" roads as well as do some four-seat saloons.

The strange wheel layout, and chassis proportions with wheelbase about 1 ft. greater than the front track, might be expected also to have an unusual effect on roadholding characteristics. The only strangeness which arises in practice is that when, in extreme conditions, the rear wheels are made to slide, the forward-mounted driver with no bonnet in front of him has slightly less warning. However, despite a rear axle without differential, sliding is infrequent even on a wet road; rigidly connected twin driving wheels are on the other hand a distinct advantage with a lively engine and a weight distribution placing no more than three and a half hundredweight on the rear. A little wheel-spin is still possible if the clutch is misused.

On a dry surface roadholding is excellent, and roll almost non-existent. Many drivers would prefer not to have such strong understeer, which may be the price of stability in straight running with a short wheelbase. Originally conceived as a left-hand drive vehicle, with a single occupant counter-balancing the engine carried over the right-hand back wheel, the Isetta in its British form understeers even more markedly when turning left.

The one feature which calls for serious criticism is the steering layout. Here, too, adaptation for the English rule of the road has had a rather adverse effect, by introducing a long, transverse drop-arm shaft which was a source of lost-motion. Apart from the spring effect of the shaft in torsion, one of the bushes on the prototype sent for test was apparently unlubricated, and had worn enough in 1,200 miles to cause some 10 deg. of play at the steering wheel. The steering is very light but not especially direct, at two and a half turns between the extremes of a lock which for so small a car is disappointing.

Even with a 30-ft. turning circle, however, parking or other exercises in a restricted space are exceptionally easy, which raises the issue of the front-opening door. The Isetta is 7 ft. 6 in. in overall length, or about 4 ft. shorter than a typical small saloon. It can therefore be backed into a much smaller longitudinal space at the kerb, and having been manoeuvred there will be plenty of space between it and the next car—only about 2 ft. is needed—for the door to be opened.

With traffic of European dimensions parked beside the kerb, it is not really

The side engine cover reveals the carburetter, but not much else. A small hatch behind the driver's seat gives access to the sparking plug, and a cover plate in the flywheel conceals the contact breaker.

possible to make use of still smaller gaps by placing the Isetta head-on to the pavement. Allowing for the overhang ahead of the front wheels this would be practicable between cars of average American size, and where head-on parking is universal the front door is most convenient. The same might well be true of many garden sheds intended to take a motorcycle and sidecar, but not wide enough for a car.

For ease of entering, the front door is unrivalled, except that with a passenger already seated the driver's entry becomes less simple. The universally jointed steering column which swings out with it fulfils the dual purpose of making the driver's seat more accessible and providing him with a convenient handle to close the door. In the interests of draught-sealing, which is very good, the door cannot be slammed, but is finally pulled to with a rather stiff, positive latch; a spring counterweight makes opening easy. An incidental of the chassis design is that, although the front panel would offer little resistance in the case of collision, it has a very strong tubular frame.

The single bench seat is wide enough to carry two adults in comfort with quite adequate space for long legs, and at a pinch a child could also be carried. Modern motorists accustomed to the comforts of interior heating are at least partially catered for by air ducts from the fins of the air-cooled cylinder, which can be arranged to direct moderately warm air to either floor-level or the windscreen. The fact is, however, that there is not a great deal of heat to spare from 300 c.c. For summer weather the Isetta in its latest form is much better provided with ventilation by sliding windows. A simple fabric opening roof is provided, while on the export model which was the subject of our test a grille in the outer door panel admits cool air to interior or windscreen. Visibility in all directions is excellent except for a severe windscreen-pillar blind spot.

Simplicity, being a characteristic of the design, is extended to servicing and maintenance, and the minimum of attention is needed. So far as the engine is concerned this is perhaps as well, for the carburetter is the only component easy to reach when the engine cover is removed. A small hatch behind the driver's seat gives access to the sparking plug.

The Isetta is unusual in a number of ways. Not the least of these is its place mid-way between full-sized car and motorcycle, with considerable appeal to owners of both. Sporting enough in performance to relieve the monotony for a motorcyclist seeking better weather protection, it is sufficiently refined for the motorist compelled to look for extreme economy. Backed by a factory in this country, it should soon be a familiar sight on British roads.

The crux of the design is the Isetta's front-opening door, with its hinged steering column which makes entry or exit very simple. Weather-proofing is good.

Specification

Engine

Cylinders	1, fan cooled
Bore	72 mm.
Stroke	73 mm.
Cubic capacity	298 c.c.
Piston area	6.3 sq. in.
Valves	Overhead (pushrod)
Compression ratio	7.0/1
Carburetter	Bing 1/24
Fuel pump	None, gravity feed
Ignition timing control	Automatic
Oil filter	Full-flow
Max. power (net)	13 b.h.p.
at	5,200 r.p.m.
Piston speed at max. b.h.p.	2,500 ft./min.

Transmission

Clutch	F. and S., single dry plate
Top gear	6.1
3rd gear	8.17
2nd gear	12.18
1st gear	23.21
Reverse	30.0
Final drive	Enclosed chain
Top gear m.p.h. at 1,000 r.p.m.	9.3
Top gear m.p.h. at 1,000 ft./min. piston speed	19.4

Chassis

Brakes	Hydraulic
Brake drum internal diameter	7.1 in.
Friction lining area	50 sq. in.
Suspension:	
Front	Dubonnet, coil springs
Rear	Rigid axle, quarter-elliptic leaf
Shock absorbers	A.T.E. telescopic hydraulic
Steering gear	B.M.W.
Tyres	4.40—10

Coachwork and Equipment

Starting handle	No
Battery mounting	Under seat
Jack	Screw
Jacking points	Front cross member and rear axle
Standard tool kit: 5 open-ended spanners, screwdriver, plug spanner, tommy bar, carburetter ring spanner, tool roll.	
Exterior lights: 2 head/side, 2 tail, 1 stop/number plate, 2 indicators.	
Number of electrical fuses	6
Direction indicators	Flashing amber, non-cancelling
Windscreen wipers	Two-blade electric, non-parking
Windscreen washers	None
Sun vizors	None
Instruments:	Speedometer with non-trip, non-decimal distance recorder
Warning lights: Dynamo charge, headlamp main beam, direction indicator.	
Locks:	
With ignition key	Ignition
With other keys	Door, engine cover
Glove lockers	None
Map pockets	One on door
Parcel shelves	One behind seat
Ashtrays	None
Cigar-lighters	None
Interior lights	None
Interior heater	Optional, fresh air type, with demister
Car radio	Optional
Extras available: Front bumper, radio, rear luggage carrier, locking petrol cap, windscreen de-mister, portable lead lamp, seat covers, additional floor mats, wheel embellishers, spare wheel, two-colour finish, windscreen washers, interior grab handles, interior luggage rack.	
Upholstery material	Cloth
Floor covering	Felt and rubber mat
Exterior colours standardized	Four, two-tone optional
Alternative body styles	None

Maintenance

Sump	3 pints, S.A.E. 20 (winter), 40 (summer)
Gearbox and differential	1 pint, S.A.E. 40
Rear axle	½ pint, S.A.E. 40
Steering gear lubricant	Grease
Chassis lubrication	By grease gun every 1,000 miles to 4 points
Ignition timing	8° b.t.d.c.
Spark plug type	Lodge HH14
Spark plug gap	0.024 in.
Contact-breaker gap	0.016 in.
Valve timing: Inlet opens 6° a.t.d.c.; inlet closes 34° a.b.d.c.; exhaust opens 34° b.b.d.c.; exhaust closes 6° b.t.d.c.	
Tappet clearances (cold):	
Inlet	0.004-0.006 in.
Exhaust	0.006-0.008 in.
Front wheel toe-in	$\frac{5}{32}$—$\frac{3}{16}$ in.
Camber angle	1° 30′
Castor angle	5°
Steering swivel pin inclination	12°
Tyre pressures:	
Front	20-22 lb.
Rear	17-18 lb.
Brake fluid	Lockheed
Battery type and capacity	12-volt, 24 amp. hr.

Isetta
VELAM

Construction entièrement française
(Licence ISO - MILANO)
SIÈGE SOCIAL
11, Rue Leroux, PARIS (16e)

USINES ET SERVICES COMMERCIAUX
1 et 3, Rue Salomon de Rothschild
SURESNES (Seine)

EDIT. G. BUNEL

CARACTÉRISTIQUES TECHNIQUES

CARROSSERIE:

Monocoque à porte frontale dégageant complètement la banquette - Pare-chocs enveloppants très robustes - Ouverture de la porte par bouton-poussoir à serrure - Intérieur soigné en simili cuir - Accoudoirs - Commandes électriques groupées au volant : avertisseur, lanternes, phares, codes, feux direction avec minuterie, contact - Compteur de vitesse au centre du volant - Chauffage efficace - Plafonnier - Vaste plage arrière pour les bagages - Roue de secours sous le siège - Visibilité totale - Toit décapotable.

GROUPE MOTOPROPULSEUR (Fixé par 6 écrous, aisément démontable):

MOTEUR: 2 temps - 2 cylindres en aluminium chromé - Chambre de combustion commune - Cylindrée totale 236 cm³ - Puissance fiscale 1 CV - Puissance réelle 9,5 CV à 4.500 tours/minute - Embiellage sur rouleaux - Refroidissement par air forcé par turbine - Graissage par adjonction au carburant de 3 % d'huile Energol Auto SAE 30 raffinée par la Société Française des Pétroles BP - Alimentation par gravité - Carburateur conçu spécialement par Solex, type 30 AH CD - Démarreur électrique - Batterie 12 volts, 24 ampères - Alternateur - Redresseur - Allumeur - Suspension du moteur par silent-blocs.

EMBRAYAGE: A disques multiples dans l'huile.

BOITE DE VITESSES: A 4 vitesses et une marche arrière, toutes synchronisées.

TRANSMISSION: Arbre de transmission constitué d'un élément tubulaire et de deux flectors - Chaîne double à rouleaux travaillant dans l'huile - Pont-arrière sans différentiel.

SUSPENSION: Arrière, par ressort semi-cantilever et amortisseurs télescopiques.

TRAIN AVANT:

Aisément démontable - Direction Gemmer à vis globique - Suspension avant par anneaux Neimann.

FREINS:

A PIED: Agissant hydrauliquement sur roues avant et mécaniquement sur transmission.
A MAIN: Par câble sur la transmission.

COTES D'ENCOMBREMENT:

Longueur hors tout	2,380 m
Largeur hors tout	1,420 m
Hauteur	1,320 m
Empattement	1,500 m
Voie avant	1,228 m
Voie arrière	0,506 m

PERFORMANCES:

Vitesse maximum: 85 km/h.
Consommation: 3,5 litres aux 100 km.

DIVERS:

Poids à vide: 320 kg (en ordre de marche avec roue de secours et outillage).
Capacité du réservoir: 15 litres dont 3 litres de réserve.
Pneus basse pression: 4,40 x 11.

PRIX:

297.000 F, plus taxe locale.

PAIEMENT:

20.000 F à la commande - 30.000 F trois mois au plus, avant la livraison - Le solde à la livraison ou par mensualités - (CONSULTEZ VOTRE CONCESSIONNAIRE).

INDIQUEZ-NOUS votre adresse pour recevoir la liste de nos Concessionnaires.

M ....
....
....

ISETTA

car comfort

at motor cycle running costs

The new 3 wheeler

the little miracl

The ISETTA is the ideal small car—cheap to buy and run, comfortable, reliable and simple to drive.
ISETTA offers you a choice of three models in the 3 wheeler range—Standard Saloon, fully equipped ready to drive away, and the PLUS and SUPER PLUS models with additional accessories already fitted.

Or you may prefer the 4 wheel ISETTA, shown on left, equipped the same as the PLUS.

This is the famous 'little miracle' that completed the gruelling 1000-mile Mille Miglia race at an average speed of 49.6 m.p.h. and on official trial covered 2,350 miles from Sicily to Sweden in 62 hours—an average of 37.9 m.p.h. *including* stops.

ISETTA cars are right hand drive

ISETTA

with a big heart and a tiny thirst

inside information

The ISETTA really is comfortable. Bench seat takes two passengers beside the driver, sitting relaxed with ample leg room.

Quarter-elliptic leaf-springs and long-stroke telescopic hydraulic shock absorber at rear, plus independently sprung front wheels give the ISETTA its wonderfully smooth riding qualities.

Built-in, draught free, weatherproof sun roof and sliding windows make it snug in winter and fresh as the dawn in summer. A highly efficient heater circulating fresh warm air is fitted in PLUS and SUPER-PLUS models or as an optional extra in the Standard Saloon.

The unrestricted all-round visibility is a joy to both driver and passengers. And you know you're secure riding in an ISETTA with its rigid tubular steel chassis frame.

EFFORTLESS DRIVING

Turn the ignition key starter and immediately the world-famous BMW 295 c.c. 4-stroke engine purrs into action, even in the coldest weather. Get-away is smooth and acceleration rapid through the four forward synchromesh gears (and one reverse). Quick, accurate reaction of the positive steering makes the ISETTA easy to manoeuvre through traffic and to control while cruising at fifty on the open road.

Easy Parking

The ISETTA is the easiest car in the world to park. You can safely drive in nose to kerb without protruding more than average car width. "Car Park Full" signs rarely apply to an ISETTA. Garage accommodation is easy—and cheap—too. But even if there's no garage you can leave an ISETTA in the open (safe whatever the weather with the air—not water—cooled engine). To protect the body work ISETTA supply a most efficient fitted coverall car mack at only £4.16.

Luggage Facilities

Very large shelf behind the seat provides plenty of room for normal luggage, shopping or picnic kit. If there's a *lot* of luggage a sturdy grid in polychromatic or Chrome finish can be fitted as an optional extra.

Polychromatic £4.10.0 plus fitting
Chrome £6.10.0 plus fitting

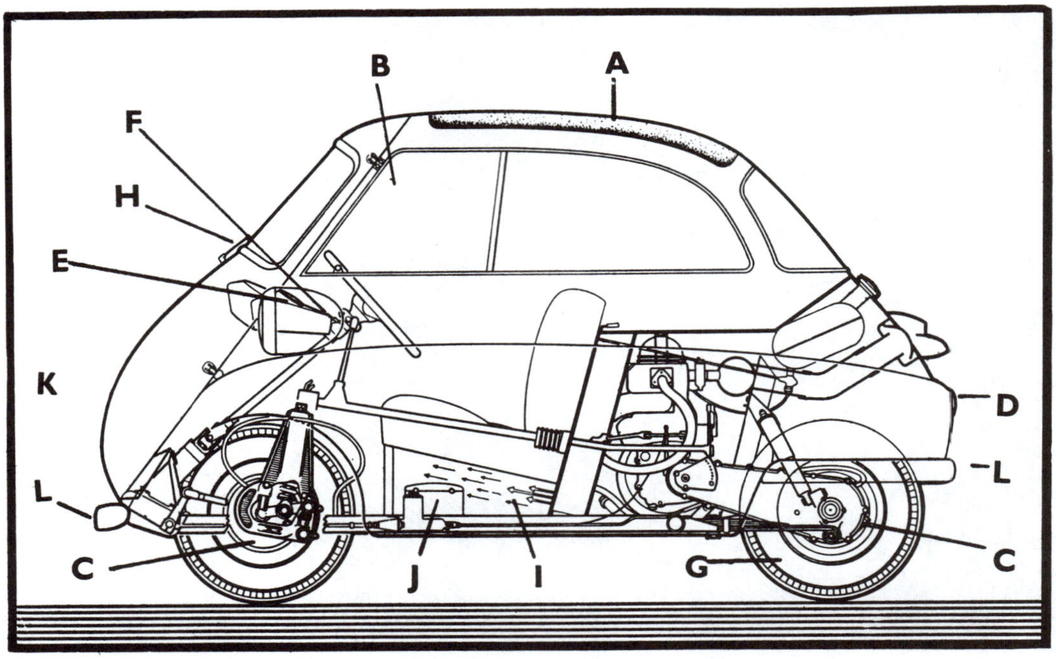

Standard Equipment in ALL Isetta 3-wheeler models

- **A** Built-in sun roof.
- **B** Sliding windows.
- **C** Hydraulic brakes to all wheels.
- **D** Twin rear lights. Illuminated number-plate and brake stop light.
- **E** Electric ignition starter. Hand dipper and trafficator controls.
- **F** 12-volt electrics, by Lucas.
- **G** Long-life Dunlop tubeless tyres.
- **H** Twin windscreen wipers.

Additional Equipment included with PLUS and SUPER-PLUS models

- **I** Built-in heater and demister.
- **J** Comprehensive tool kit under seat.
- **K** Large glove and papers pocket in front door panel.
- **L** Front and rear bumpers.

And the SUPER-PLUS model also includes spare wheel and fitted seat belt in driver's position.
(Additional seat belts by "Britax" £4.4.0. Plus fitting)

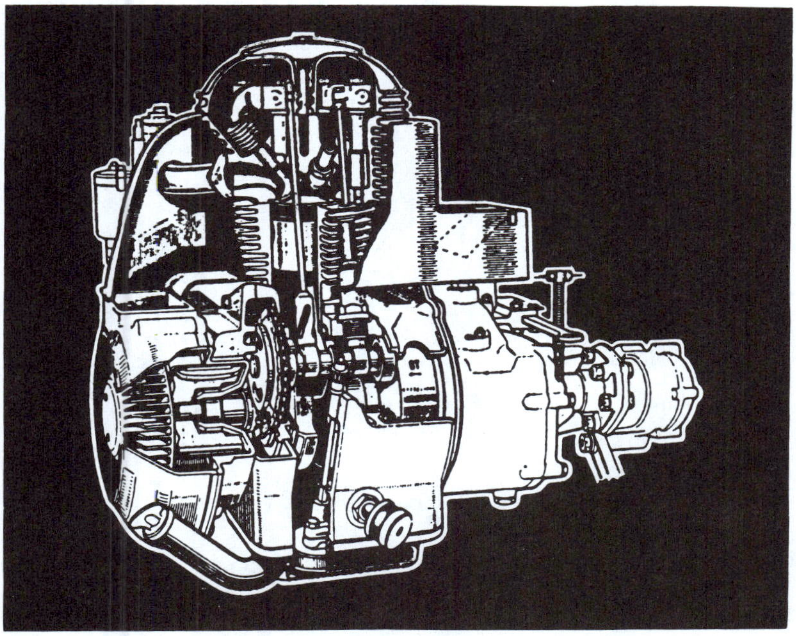

Specification

ENGINE: BMW fan-cooled single-cylinder 4-stroke sq. engine.
Bore....72 mm (2.83 in)
Stroke....73 mm (2.87 in)
Capacity....295 ccm (18.30 cu. in.)
Compression....7.0 to 1
Power....13 b.h.p.
Valves....Overhead valves, push-rod operated.
Lubrication....Circulation system with gear type pump.
Starter....12-volt dynamo starter.
Electrical Equipment....12-volt/130 watt generator.
Engine Position...Right-hand side, transverse behind the seat.
Clutch....Single plate dry clutch.

TRANSMISSION: BMW four forward speeds and reverse (optional) gearbox attached to the engine. Final Drive: Resilient mounted drive shaft and totally enclosed fully-adjustable chain drive in oil-bath.

CHASSIS FRAME:....Rigid tubular steel chassis frame.
WHEEL SUSPENSION: Independent front-wheel suspension; swinging arms, coil springs and shock absorbers. Rear-wheel suspension; quarter-elliptic leaf springs and telescopic long-stroke hydraulic shock absorbers.
WHEELS:....By Dunlop
TYRES:....Long-life Dunlop tubeless tyres. Size: 4.80-10.
BRAKES: ...Hydraulic brakes by Girling to all wheels. Brake diameter 7 ins.
DIMENSIONS: Track, front....47.2 in.
Wheelbase....58.0 in.
Overall length....89.9 in.
Overall height....52.6 in. (unladen)
Overall width....55.35 in.

WEIGHT:....Kerb weight approx. 810 lb.
Carrying capacity approx. 507 lb.
ROAD PERFORMANCE: Maximum speed; 52 m.p.h.
Climbing ability; First gear 1 in 3.
Fuel Consumption; 83 m.p.g. at constant 30m.p.h.
FUEL TANK CAPACITY: 2.8 Imp. gals. with reserve fuel supply of .65 Imp. gal
COLOURS: STANDARD MODEL; Imperial Red, Riviera Blue.
PLUS and SUPER PLUS MODELS;
Imperial Red, Cardinal Red, Riviera Blue, Ice Blue, Azure Blue, Highway Yellow, Ivory and Light Green.
Note: Isetta of Great Britain Limited reserve the right to introduce modifications in the design of the ISETTA without notice.

Maintenance and Spares

ISETTA means trouble-free economy motoring and that applies equally to after sales service. In planning and designing the ISETTA one of the foremost considerations was to keep maintenance costs down to a minimum. Cheap-to-service engine, long-life tubeless tyres and thoroughly tested reliable components are symptomatic of this. There is a world-wide ISETTA Service Organisation, with works trained engineers backed by a good spares service if ever you need it.

And there's a whole range of attractive accessories you can add to your ISETTA whenever you like. Have a word with your dealer.

B M W ISETTA "300"

1957/1958 MODEL (SLIDING SIDE WINDOWS)
METAL INSERT IN CABRIOLET

B M W ISETTA "300"

1957/1958 MODEL (SLIDING SIDE WINDOWS)
METAL INSERT IN CABRIOLET

METAM

A Minor Industrial

which would provide new goods for export, and simultaneously help to solve a local labour problem.

Part of the British Railways locomotive works at Brighton, adjoining the main station, was going out of commission. In March, 1957, the last repair jobs from that bay were completed, and the last new steam locomotive from the Brighton works dispatched. The contractors moved in the same week; assembly lines, welding jigs and electrical plant followed. In three weeks' time, 70-ton railway engines will have given way to 7-cwt. cars, assembled in the same place by many of the same hands, which proves the adaptability of men as well as buildings.

The plan, however, involved more than just taking over floor space. The old locomotive works is at the top of a precipitous hill. It is not, and short of tunnelling cannot be, reached by road, but there is an adage about making a virtue out of necessity. If parts have to finish their journey by rail, why should they not start that way, especially as the components arriving from Germany must do so in any case? And for the delivery of complete cars there is a great deal to be said for the railway.

A design for this new and highly original factory was produced by B.M.W. for the British company, and its

IN a quiet way the Isetta motocoupé, which originated in Italy, acquired the sobriquet of Easter Egg in France and was taken up for large-scale production by one of the biggest motorcycle manufacturers in Germany, has been doing very well. The Bavarian works now turns out 220 cars a day, the total running to over 65,000, and of these 1,000 have come to Britain under the stimulus of the shortage and ever-increasing price of petrol. Some months ago it was revealed that plans were afoot to produce the car here, and the imminent appearance of British-built Isettas was announced in London yesterday.

Anyone without a particularly inquiring frame of mind might well have seen nothing odd in the fact that Dunsfold Tools, Ltd., had as their address the Locomotive Works, Brighton. Plenty of companies have unusual addresses. Add to it the fact that Dunsfold Tools, Ltd., has now become Isetta of Great Britain, Ltd., building a miniature car in a factory whose only access to the road is down a flight of something over 100 steps, and you have the beginnings of a story.

First Off

The end of the first chapter and the start of the next are due on April 23, when the first British Isetta comes off the line; the true beginning was a bare nine months ago. The import of cars into many parts of the world—including countries like Canada with hard currency to exchange for them—is easier and cheaper when the cars come from Britain, which is good business for the B.M.W. concern which can supply components and draw licence fees. For Britain it is equally good business: a small outlay in deutschmarks for a large return in dollars, which is the kind of transaction to impress favourably the Board of Trade. In July of last year both the Board and the Ministry of Labour were impressed by a proposal

Heavy stuff in the Brighton locomotive works two months ago. In three weeks time the same lines will be starting production of 50 British Isettas a day.

ORPHOSIS

Revolution at Brighton

essence is shown in a diagram on this page. Of the three railways tracks only the middle one was left, for supply and delivery, while inspection pits beneath the side tracks were put to good use for the installation of endless conveyors. From the raw pressings received from Bavaria, and welded on jigs proof-tested at the German factory, the bodies are built on one line, pass into and out of a modern paint shop in a loop, and arrive at a point half-way down the opposite line on which the chassis frames have taken shape. At the end of the final assembly, 25 cars arrive at the top of a loading ramp which, apart from spaces beside the assembly lines themselves, is the only "store" in the building.

Weekly Routine

The production programme looks, naturally enough, like a railway timetable. The Brighton goods depot accumulates during the week a trainload of parts sufficient for 250 complete Isettas; engines, transmissions and body panels from Germany; electrical gear, brakes, tyres, suspension and other parts from the Midlands; small metal components from Shoreham and other local sources. On Saturday morning a week's supplies are unloaded from the central track to the delivery points in the factory. At 7.30 a.m. each day from Monday to Friday, and again during the lunch-time break, the shunting engine withdraws eight flat wagons carrying three Isettas apiece, and leaves eight empty wagons in their place. Sixteen wagons travel nightly with the express freight train from Brighton to London, and thence to anywhere in the country. The last link of the transport chain can be the dealer or, if he chooses, the customer himself, who will be at liberty to collect from his local railway station a car with a total distance on its recorder of approximately 25 yards.

(Lower picture): Convertible for the Canadian market. The same body style with normal lights, bumpers and right-hand steering will be available in Britain.

For the first few weeks the whole production effort will be concentrated on an order of 1,000 cars for Canada. Left- or right-hand drive presents no problem in assembly of the Isetta, and production will be flexible within a range of six models which were shown yesterday to the Press and to representatives of the United Kingdom dealers. Two are the Canadian cars, with left-hand steering, heavier bumpers and electrical systems to suit local conditions. The mainstay of production for the British market will be a cabriolet and a saloon (a road test report on the latter appeared in *The Motor* of March 13), and the most important item in yesterday's announcement was a reduction in the price of the standard saloon to the remarkable figure of £359 17s. 6d. with P.T. For a four-wheeled vehicle bearing the full weight of 60% purchase tax this is a step in the direction of real economy motoring. The lower rate applied to commercial vehicles brings down to £299 17s. 6d. and £307 2s. respectively the prices of the last two models available, a miniature pick-up and a "lorry."

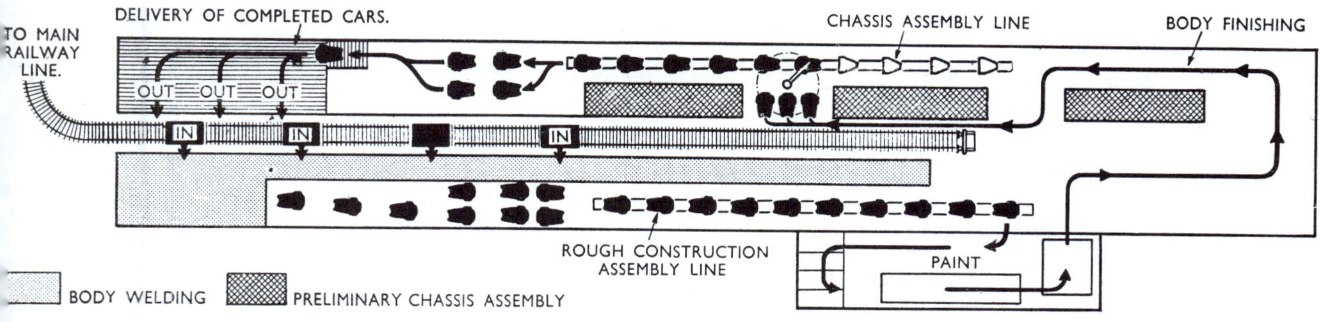

Running like a train, the factory will be supplied by rail once a week, and cleared of completed cars twice daily by the same means.

ANNOUNCING

THE BRITISH-MANUFACTURED

Isetta

Standard Saloon
£239 . 0 . 4
+£120.17.2 P.T.

Pick-up version—Fabric Top £260.7.2+£39.10.10 P.T.
Solid Top £267.11.2+£39.10.10 P.T.

ISETTA OF GREAT BRITAIN LIMITED, BRIGHTON, 1

The World's cheapest car to buy and run!

THE AMAZING *Isetta*

FOUR-WHEEL SAFETY WITH REAR TWO-WHEEL DRIVE

Most people are aware that four wheels are more stable and hold the road better than three. The Isetta has this extra four-wheel safety with the added advantage of two-wheel drive, giving almost the performance of a cross-country vehicle.

PERFORMANCE AND RUNNING ECONOMY

The Isetta will cruise "for ever" at 50 m.p.h., and at this speed it has a petrol consumption of up to 70 miles per gallon. In a recent test, it was found to have a consumption of 93 miles per gallon at a steady speed of 30 m.p.h. Moreover, it is exceptionally economical and simple to run in other ways. An air-cooled engine: one sparking-plug only: a sump capacity of only 3 pints: only 6 greasing points—they speak for themselves. With four forward synchromesh speeds (and one reverse gear) the Isetta has remarkably quick acceleration. It is fitted with Dunlop tubeless tyres, with a life of 24,000 to 28,000 miles on the front and 35,000 to 40,000 on the back.

COMFORT

Isetta comfort sets a new standard among small cars. Its bench-type seat will hold three medium-sized persons or two large, giving plenty of unobstructed leg-room. With quarter elliptic leaf springs and two long-stroke telescopic hydraulic shock-absorbers at rear, and independently sprung front wheels, it gives an exceptionally smooth ride. The built-in draught-free and rain-proof "sunshine" roof can be folded right back, and there are sliding windows to give added ventilation. There is unrestricted all-round visibility, and the Isetta's "finish" is superb. Access is from the front; the door closes silently, and is fitted with a fingertip safety lock.

FROM THE DRIVER'S VIEWPOINT

Turn the Isetta's ignition-key starter and there is an immediate response from the world-famous B.M.W. 295cc. 4-stroke engine—even in the very coldest weather. This air-cooled, overhead-valve engine is acknowledged to be one of the most robust and reliable in existence. Steering (there is a choice of right- or left-hand drive) is positive and accurate with quick reaction to parking requirements.

The Lucas electrics are 12-volt, and the headlamps are exceptionally powerful. Instruments are by Smiths.

EASY PARKING

Since its length is no more than average car width, it can be placed nose-on to the kerb, and is therefore the easiest car in the world to park.

PLENTY OF ROOM FOR LUGGAGE

A very large shelf behind the seat gives plenty of room for luggage or shopping, and in addition a luggage grid can be fitted as an optional extra.

MAINTENANCE AND SPARES

In designing the Isetta, particular attention has been given to keeping costs of maintenance down to rock-bottom. The long-life tubeless tyres and the fully tested, exceptionally cheap-to-service engine are symptomatic of this principle. Good spares service is assured, wherever you go, by the Isetta world-wide Service Organisation with its works-trained engineers.

PASS AND JOYCE LTD

ENGLAND'S LARGEST ISETTA DISTRIBUTORS

184-188 Great Portland Street, London, W.1. Tel: MUSeum 1001

Also

27 Peter Street, Manchester
Tel: Deansgate 6157

Bourne Court, The Square, Bournemouth
Tel: Bournemouth 3562

BMW Isetta Specifications

Engine	BMW fan-cooled single cylinder four-stroke engine
Bore	72 mm
Stroke	73 mm
Capacity	295 ccm
Compression	7.0 to 1
Power	13 hp
Valves	over-head valves, push rod operated
Lubrication	Circulating system with gear type pump
Starter	12 Volt starter
Electr. Equipment	12 Volt/130 Watt Generator
Engine Position	Right hand side, transverse behind the seat
Clutch	Single plate dry clutch
Transmission	BMW four forward speeds and reverse gearbox attached to the engine
Final drive	Resilient mounted drive shaft and totally enclosed, fully adjustable chain drive in oil-bath
Chassis frame	Rigid tubular chassis frame
Wheel Suspension	Independent front wheel suspension, swinging arms, coil springs and hydraulic shock absorbers. Rear wheel suspension, quarter elliptic leaf springs and telescopic hydraulic shock absorbers.
Wheels	Split-rim disc wheels. Spare wheel with tyre
Tyre Size	4.80—10
Steering	Steering wheel, reduction steering gear, smallest turning circle approx. 24 feet
Brakes	Four wheel hydraulic brakes. Brake diameter 7 in. Total brake lining area 49.9 sq. in.

Dimensions:			
Track, front	47.2 in.	Overall length	89.9 in.
Track, rear	20.4 in.	Overall width	54.3 in.
Wheelbase	58 in.	Overall height (unladen)	25.6 in.

Weight:	
Kerb weight	approx. 770 lbs.
Carrying capacity	507 lbs.
Road Performance:	
Max. Speed	52.8 miles per hour
Climbing ability	First gear 1 in 3
Fuel consumption	approx. 76 miles to the Imperial Gallon

Modification to details and specification without reserve
Body design licensed by Iso Spa. Milan/Italy

Over One Hundred Thousand Satisfied Owners

1957

Isetta Motors (Canada) Limited
Temporary Office: The Park Plaza Hotel, Suite 1037, Avenue Road and Bloor Street, Toronto, Ont.
WALNUT 2-0243
DEALER FRANCHISE INQUIRIES INVITED

INTRODUCING

a new conception in vehicle design

Large sized and ideally comfortable front door

"If European car makers can produce an auto efficient enough and cheap enough for Europe's workers, a gigantic new market can be opened up." These were the lines quoted by a popular American Economic Review

Meanwhile the appearance and outstanding performance of the BMW Isetta has favorably influenced the problem of generally motorizing Europe's population.

No parking problems - smallest possible parking space

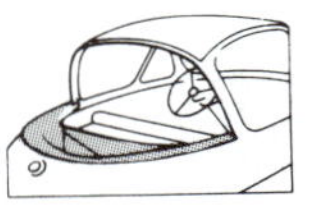

Besides the standard design with canvas hood the BMW Isetta can also be supplied with a convertible hood - in a few seconds the hood can be opened or closed

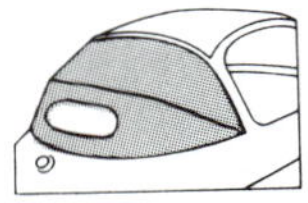

The well-known BMW four stroke engine

The latest product of the well-known German company Bayerische Motoren Werke AG Munich, in short called BMW, who has acquired a great name during several decades by its activities in the field of air-craft-engine, automobile and motorcycle construction, proved to be the ideal solution in modern design of midget cars. The BMW Isetta is an entirely new type of vehicle. Due to its special features in design, its low price and its economy, it is predestinated to be the all-purpose vehicle for everybody. Many thousands of Europeans have enthusiastically welcomed it;

we are sure that it will also arouse a vivid interest in America, especially on account of the traffic problem which demands new ideas in vehicle construction and which is in bad want of versatile, room saving and economical midget cars: the BMW Isetta is indeed apt to fulfil all these requirements.

After its big success in Europe the BMW Isetta is now being introduced in Canada and these are its special features of design.

- The spacious and aerodynamically favorable body has been developed from the somewhat modified shape of an egg. It offers sufficient space for two grown up persons and one child, as well as luggage.

Safety by rear wheel transmission without differential

- Large windows offer an allround view, the practical canvas hood can be opened in a second.
- The ideally comfortable front door has replaced the traditional and expensive side doors.
- Four wheels guarantee safe driving in every situation and due to the fact that the rear wheels do not have a differential, the BMW Isetta can cope even with the most difficult road surface.
- The responsive BMW four stroke engine and its easy steering make the BMW Isetta an ideal means of communication in the crowded traffic situations of our days.
- Every single technical detail is a proof of BMW precision and craftsmanship the finishing characteristics of which follow the lines distinguishing BMW cars.

We are sure you want to know more details, therefore please contact our BMW Isetta sole agents in Canada, who will take pleasure in offering you a test drive.

Isetta
I-SETTA

The EGG and I

New British-built Isetta is a real surprise packet, reports Harold Dvoretsky after testing it in town and country

WELL, it's happened. The little mechanised eggs which originated in Italy and spread like a rash all over Europe in the past two years are now invading England.

And it's no use talking of throwing them back into the sea, for they've established a firm beach-head in an ex-locomotive factory at Brighton and are streaming out in ever-increasing quantities.

As long as they kept out of this tight little island, I gave little thought to the Isettas. I'd watched them take over Italy, then France, read a little incredulously of their performance in the 1955 Mille Miglia (all four finished and the fastest averaged 49.6 for 998 miles), noted that Germany's BMW factory alone had hatched 70,000 of them in two years.

But all this was happening on the Continent, where people like their cars small and peculiar. They'd never take on in England, I thought, or in Australia . . .

CLIP OFF a bit of "eggshell" and there's the little engine.

UNORTHODOX but neat, and comfortable inside for driver and passenger. Dvoretsky's wife went on the test to give the woman's angle on the car.

"It's the first British-built prototype," said Ronnie. "Hop in. I'll take you round for a familiarisation trip." (He was in the R.A.F. during the war, and a B.O.A.C. pilot for some years afterwards — hence the Air Force jargon.)

Ronnie opened the front door and I piled in beside him. Entrance was easy, anyway, I thought. A touch on the key-type Siba dynastarter and the little air-cooled engine came swiftly to life. In no time at all we were tearing about Brighton streets.

I was just beginning to take notice of things: the passenger side of the bench seat was comfortable, controls looked simple, there was no feeling of roll, no pitching on sudden braking . . .

"Yours now, said Ronnie, pulling into the kerb.

Three stops and starts from rest and I'd mastered the unfamiliar, tiny gearshift at my right hand. Later

I was wrong. In the past few months, Isettas started popping up in London streets — first singly, then in dozens. Before long, proud owners were earbashing me about the roadholding, the performance, the economy of their little eggs . . . It was worse than talking to a bloke with a VW.

Then Ronnie Ashley rang to say he'd left his job as head of Armstrong-Siddeley's car division to start a BMW Isetta factory at Brighton, with a production capacity of 500 a week.

This was the last straw. "Okay," said I. "When can I test one?"

The Egg and Ronnie

That's how I found myself at Brighton Station one fine morning, gazing at a little bubble-shaped four-wheeler parked alongside the platform and listening to Ronnie's enthusiastic patter:

"There will be four models — a saloon, a cabriolet, a delivery van, a light goods van . . . Canada has already ordered 4000, mostly cabriolets . . . It'll sell at £239 sterling basic — cheaper than the German-made job . . . We're planning a five-seater model, too . . ."

Then Ronnie dropped his bombshell: "This is the cheap car for Australia — oodles of performance, economy and reliability!"

I muttered something about bad roads, and spent the next minutes listening to how Ronnie runs his Isetta cross-country on his 500-acre farm. "That I'd like to see," I said. "You will," said Ronnie, "you will."

He whipped me through the factory, next door to the station, showed me the £70,000 assembly line which was to start rolling in a week's time — then out to the car again.

COCKPIT seen through sun-roof of prototype. Dip-switch is left of steering column, trafficator switch right. Production cars will have trafficator switch in centre of front door, dip-switch alongside gear lever, heater and demister. Warning lights are for trafficators, ignition and high beam.

SPECIFICATIONS

ENGINE: 1-cylinder, 4-stroke, o.h.v.; bore 72mm., stroke 73mm., capacity 298 c.c., compression ratio 7 to 1; maximum b.h.p. 13 at 5200 r.p.m.; gravity feed, pump lubrication; cooling by ducted fan; 12v. ignition.

TRANSMISSION: Single dry-plate clutch, 4-speed and reverse gearbox; final drive by chain in oil-bath.

SUSPENSION: Front independent, by swinging arms and coil springs; quarter-elliptics at rear; telescopic shock-absorbers all round.

STEERING: Burman reduction gear, 2½ turns lock-to-lock.

WHEELS: Dunlop discs, with 4.80 by 10in. tubeless tyres.

BRAKES: Hydraulic on all wheels; 49.9 sq. in. lining area.

CONSTRUCTION: Tubular-steel frame incorporated with steel body.

DIMENSIONS: Wheelbase 58in.; track, front 47.2in., rear 20.4in.; length 89.9in., width 54.3in., height 52.6in.; road clearance 7½in.

KERB WEIGHT: 7cwt.; load 500lb.

FUEL TANK: 2.8 gallons.

PERFORMANCE

CONDITIONS: Every type of road in 194-mile test. Two occupants, standard petrol.

MAXIMUM SPEED: 52 m.p.h.

ACCELERATION: 0-30, 11.5s.; 0-40, 23s.; 3rd, 10-30, 13s.; top, 20-40, 28s.

CONSUMPTION: 60 m.p.g. overall.

I was to grow very fond of it, despite some minor idiosyncrasies.

"Turn left, then left again at the traffic lights, and you're on the road to London. And don't let anyone put it over on you," said Ronnie.

On My Lonesome

Ronnie needn't have worried. After the first couple of traffic lights I was flying off ahead of most cars — much to my own and other drivers' surprise.

Exactly 100 minutes and 54 miles later I pulled up outside my Fleet Street office.

For most of the journey to London Id travelled with the needle beyond the 55 m.p.h. mark. The speedo in the prototype was German and somewhat optimistic — by 6 m.p.h. at that speed; British production models will have a Smith's instrument. Nevertheless, the speed and the average — considering the thick traffic — were very good.

The original Isettas from Italy had a two - cylinder engine. BMW changed this to a 245 c.c. single-cylinder job and later raised capacity to 298 c.c., boosting the little egg's performance.

At roundabouts and on 90-degree corners the Isetta raced round safely. The British-built machines have coil springing at the front and a rigid axle with quarter-elliptic leaf springs at the rear, plus telescopic hydraulic shock-absorbers all round. This suspension irons out the bumps remarkably well and keeps the craft on a pretty even keel.

I wondered what effect the car's unusual tracking would have (rear track is 27 inches narrower than the front), but road-holding was perfectly normal. When I deliberately induced a tail-slide, the reaction worried me a trifle. Not that it seemed dangerous — far from it — but I got less than the usual warning of the coming slide, because there was no bonnet in front of me.

However, the Isetta seldom slides, even on wet roads, despite the fact that it has no differential. And there's about 3½ cwt. over the back wheels, which is a distinct advantage with that lively engine.

That's all I was able to note on the first day, being very much occupied with dodging the traffic in a still unfamiliar vehicle. The real test was yet to come.

The Egg and Angela

When I got home that night, after a 15-mile drive from London, Angela watched me drive up. She was chuckling. "It doesn't look as bad as I thought," she said. "In fact, I rather like it."

Angela is my wife. Except during

DOOR opens wide, takes steering wheel with it. Ornamental bumpers are extras; heavier tubular ones are also available. Headlights are powerful.

CABRIOLET model is the one favored in Canadian orders, will probably be among cars sent to test Australian market. BELOW: Tiny "delivery van" has a steel locker neatly fitted into open back of cabriolet-type body.

THE EGG AND I

the 12 months she spent in hospital, Angela has sat in the front seat of most cars I've tested over the past four years. She still doesn't know what makes them tick — but on such things as comfort, finish and appearance she's a deadly accurate critic. When she said she liked the looks of the Isetta, it meant that a lot of other people would like them, too.

Since the Isetta was being plugged as a good second car for the family, I wanted a woman's angle on it. So I asked Angela if she'd like to double-yolk with me while I put the little egg through its paces, and she said yes, very much. Another good omen for Ronnie Ashley's venture.

In the next two days Angela and I took that little car everywhere — and anywhere. We found the bench seat wide and comfortable enough for two adults, with perhaps a child in between at a squeeze. Ample leg-room for driver and passenger, excellent visibility, apart from thickish windscreen pillars.

On rough potholed surfaces, where there's no avoiding the bumps because of the wheel layout, the ride was solid but not really uncomfortable. Despite a wheelbase of only five feet, the Isetta went over some frightful roads without giving us too bad a ride and certainly no qualms as to stability. It rode as well as some four-seat light cars I'd taken over the same stretches.

Originally the car was designed with the engine to be at the opposite side to the driver. It stayed there when the steering was shifted to the right in the British version, with the result that the already pronounced understeer was slightly increased. But the steering is so light and fairly direct (2½ turns lock-to-lock) that I soon became accustomed to it.

When we put the stopwatch on her, the Isetta went from 0 to 30 m.p.h. in 11½ seconds and reached 40 in 23. True maximum speed was 52 m.p.h., and a check-up at the end of the test showed a consumption of exactly 60 m.p.g. overall, for a distance of 194 miles. The makers claim 58 to 84 m.p.g., depending on load, speed and conditions. Standard petrol was used, as recommended.

On normal cruising I would expect the 2.8-gallon tank to give a range of about 200 miles, which would be a fair distance to go in one stretch.

Returning home at the end of the second day, we sat down and exchanged impressions.

Angela liked the easy entry ("I'd go even in evening dress," she said), the good visibility and the opening roof, which didn't create a draught if the sliding side-windows were closed. She liked the space behind the seat for handbag or parcels and pointed out that heavy shopping bags could be placed on the flat, wide floor between driver and passenger. On the debit side she put the noise from the engine; it was louder than average, although not enough to be annoying.

I liked the delightful four-speed gearbox which will take you to 40 in third and pick up nicely in that gear from 15 m.p.h. or even less; the sturdy little engine that churns out 13 b.h.p. at 5200 r.p.m.; the brakes, which, I'm told, Girling are making even better for production models. And I liked the easy parking—the Isetta is only 7ft. 9in. long, and you can tuck it in anywhere. You can even drive it straight into the kerb, and if you angle it slightly, the tail won't protrude any farther than a normally parked big American car.

About the only real criticism I have of the car is that it could have a little more protection in front. Bumper bars are available as an extra, but these are of the upright type. It shouldn't be too difficult to design two strong bars to go across the front door, giving protection in the unlikely event of a head-on collision.

On final summing-up, that night, our verdict was that the Isetta is almost the perfect town car. It's performance is most respectable, it's nippy and eel-like in traffic, and parking is no problem. As a second car to take the Missus shopping, the kids to the beach, or Dad to the office, it certainly fills the bill. And for those whose pockets are lean, it could very well fill an empty garage.

But we still couldn't quite see it in the Australian bush — until Ronnie Ashley came on the scene again.

The Scramble Test

He hadn't forgotten he'd promised to show me what he does with the Isetta on his Surrey farm, so thither I went one Sunday.

The Canadian who has bought 4000 of Ronnie's little eggs was also there. The three of us jammed ourselves into the same car I'd tested, and off we went.

We plunged and bounced our way up and down 1-in-3 slopes of rough hillsides and over a two-mile circuit of potholed, rut-creased tracks. The Isetta pulled through some stuff that I thought might have foxed a jeep.

It was a hair-raising experience—until you saw that no matter what Ronnie did with the egg, it didn't let him down. I came out of it with a new respect for the Isetta, and when Ronnie said he was trying for an Australian import licence, I didn't attempt to dissuade him (as I'd intended to do previously). Instead, I started explaining ho he could keep costs down by doing most of the assembling in Australia.

Now he's planning to send half-a-dozen Isettas "down under," to show the Aussies what they can do. And when you see this new British-built version in action, I think you'll be as impressed as I was. • • •

ISETTA

WHEELBASE: 58
WEIGHT: 770
ENGINE: 1 CYL., OHV
HORSEPOWER: 13
PASSENGERS: 2
MODEL: 1-DOOR COUPE
PRICE AT N. Y. $1,048

AS YOU drive down the street in your Isetta you've got to be prepared for the stares and amused smiles of everyone in sight. There's some justification for this sort of attention, too, because the two-passenger German import looks for all the world like a three-wheeled phone booth.

Swing open the single door (actually the front panel which includes windshield and wiper, steering wheel instrument panel and switches) and walk in, however, and you find that you are in a roomy, two-passenger automobile with conventional brake, clutch and accelerator pedals.

There's plenty of head and elbow room, a comfortable seat back and cushion, a wide windshield and a parcel shelf behind the seat. Turn the ignition key starter and you fire up a single cylinder, air-cooled engine mounted back of the seat and driving the short rear axle by means of a chain (there's no differential).

The transmission is a four-speed, constant-mesh type with engagement by sliding sprag clutches. Gearshift control is at the driver's left and surprisingly easy to operate.

As you drive the car, you find that despite its narrow rear tread, wheels are 20 inches apart, and 58-inch wheelbase, that the Isetta is very stable and well-behaved even on rough pavement.

The 13-horsepower engine is noisy and requires frequent use of the shift lever to keep up with the stop-and-go surge of U.S. traffic. However, keep up it does, and while the rest of the cars are gulping gas at the rate of a gallon every 10 miles, the Isetta is logging a 60 mile to the gallon average.

The Isetta feels safe and reasonably comfortable at 45-50 mph on average roads but we suspect that most of the Isettas sold will be wriggling through traffic at a 30 mph average.

Summing up: An ideal transportation car for the person whose driving consists mainly of short hauls in city or suburb. Isetta with its one cylinder and 700-lb. weight is the ultimate economy car.

'NEWSGRAM'

SALES & SERVICE

Isetta

ISETTA MOTORS of CANADA

Sales Representative:

$1098

F.O.B. Toronto

75

Miles per Gallon

1172 BAY STREET **TORONTO** **WAlnut 2-1106**

THE FABULOUS ISETTA 300

$1098

F.O.B. Toronto

License Extra

$366.00

DOWN·

$9.00

PER WEEK

* * *

75

Miles per Gallon

EQUIPMENT

Built-in sun-roof.

Built-in heater.

Sliding windows.

Brakes on all four wheels.

Twin rear lights. Illuminated number-plate and brake stop light.

Comprehensive tool kit under seat.

Electric ignition starter. Hand dipper & trafficator controls.

12-volt electrics by Lucas.

Long-life Dunlop tubeless tyres. All wheels interchangeable.

NOTE
The manufacturers of the British-built B.M.W. Isetta reserve the right to introduce modifications in the design of the Isetta without previous warning.

ENGINE B.M.W. *fan-cooled single cylinder four-stroke square engine*

Bore	*72 mm. (28.3 in.)*
Stroke	*73 mm. (28.7 in.)*
Capacity	*295 c.cm. (18.30 cu. in.)*
Compression	*7.0 to 1*
Power	*13 b.h.p.*
Valves	*Overhead valves, push-rod operated.*
Lubrication	*Circulation system with gear type pump.*
Starter	*12-volt dynamo starter.*
Electrical Equipment	*12-volt/130-Watt Generator.*
Engine Position	*Right-hand side, transverse behind the seat.*
Clutch	*Single plate, dry clutch.*

TRANSMISSION: *B.M.W. four forward speeds and reverse gearbox attached to the engine.*
Final drive: Resilient mounted drive shaft and totally enclosed, fully adjustable chain drive in oil-bath.

CHASSIS FRAME: *Rigid tubular chassis frame by Rubery Owen.*

WHEEL SUSPENSION: *Independent front wheel suspension by Rubery Owen and Girling, swinging arms, coil springs and shock absorbers. Rear wheel suspension, quarter elliptic leaf springs and telescopic long stroke hydraulic shock absorbers.*

WHEELS: *By Dunlop. Disc type with spring on hub-caps.*

TYRE SIZE: *4.80-10 Dunlop Tubeless.*

STEERING: *Steering wheel, reduction steering gear by Burman.*

BRAKES: *Four wheel hydraulic brakes by Girling. Brake diameter 7 in. Total brake lining area 49.9 sq. in.*

DIMENSION

Track, front	*47.2 ins. Overall length 89.9 in.*
Track, rear	*20.4 in. Overall width 54.3 in.*
Wheelbase	*58 in. Overall height 52.6 in. (unladen)*

WEIGHT

Kerb weight	*approx. 770 lbs.*
Carrying capacity	*507 lbs.*

ROAD PERFORMANCE

Max. Speed	*62 miles per hour.*
Climbing ability	*First gear, 1 in 3.*
Fuel consumption	*Town and Country, 75 miles per gallon.*

EXTRAS: *Comprehensive range of extras available. See supplementary list.*

OTHER MODELS: *There is also a pick-up version of the B.M.W. Isetta. Ask for details from your nearest Isetta agents.*

A single door opens on the rear seats, which are comfortable and easily reached.

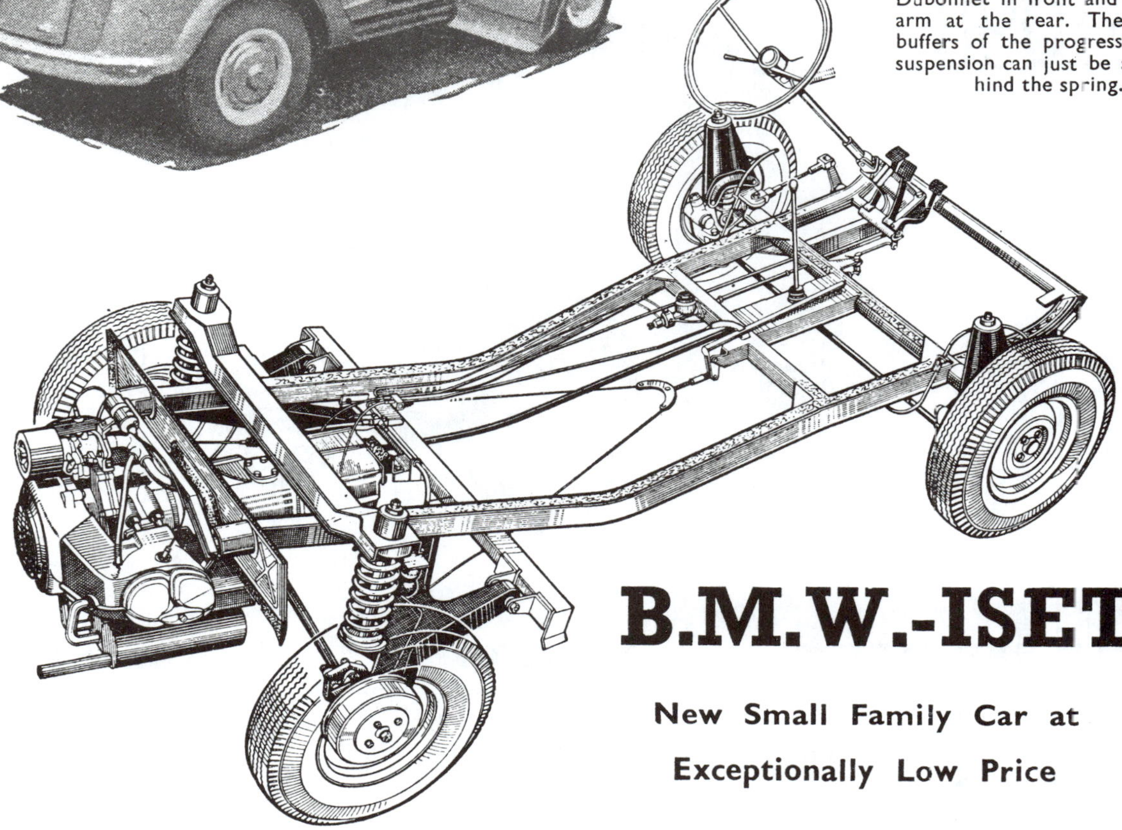

Simple and strong, the new B.M.W. chassis has all-independent coil spring suspension, Dubonnet in front and trailing-arm at the rear. The rubber buffers of the progressive rear suspension can just be seen behind the spring.

B.M.W.-ISETTA

New Small Family Car at Exceptionally Low Price

A NEW and most serious challenger in the world small-car market was revealed in Germany last week by the B.M.W. concern, where a four-seat development of the successful Isetta Motocoupé is due to go into production during November. The British-licensed version may be expected from the Brighton works of Isetta of Great Britain, Ltd., next spring.

Briefly, the achievement of the Bayerische Motoren Werke is a vehicle capable of carrying four adults in reasonable space and considerable comfort, or two occupants and a great deal of luggage, within an overall length of 9 ft. 6 in., and at a price which in this country will be £479 17s. including tax. Moreover, brief acquaintance with the car on roads near Munich shows that it is both quiet and lively, and should be genuinely economical.

The B.M.W. 600, to give it its German name, is a logical extension of the scheme pioneered by Iso of Italy, and greatly improved when the Isetta was taken up for large-scale manufacture by B.M.W. Fierce commercial competition and the unchangeable nature of the human frame have stimulated

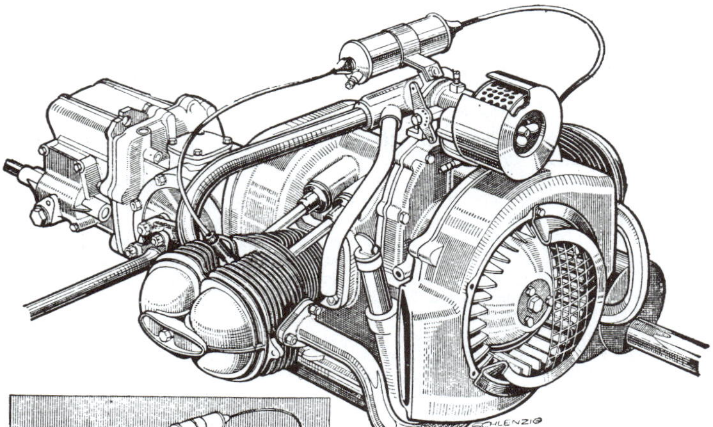

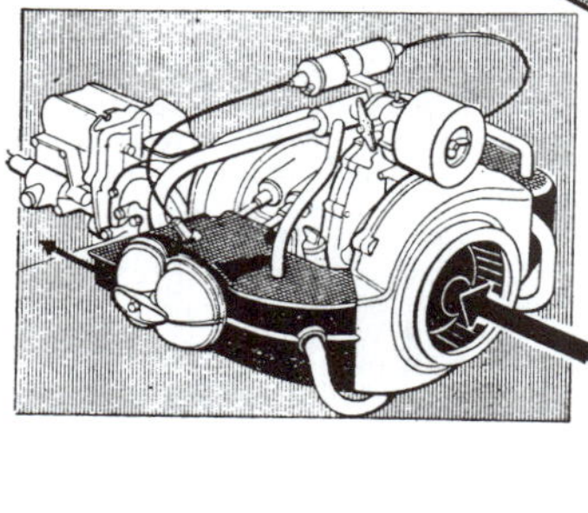

The flat-twin engine is a "slogging" version of the old motorcycle unit, with ducted fan cooling and a single carburetter. Pipes leading upward from the ducts take warm air to the inlet pipe hot-spot. Interior heating is by air from the cylinders, while a smaller supply can be passed over the exhaust and directed to the windscreen for demisting.

Small outside and large inside, the 600 carries four people within its 9 ft. 6 in. length.

600

Detail changes from the Isetta Motocoupé are a central gear lever and handbrake, and steering column with two universal joints.

original design in the field of small cars perhaps more than in any other. Experience had shown the front-opening door to have several advantages with few snags, while a compact engine of suitable size was ready and waiting for modification, so that the launching of a model to fill the gap between two-seat runabout and luxury saloon was by no means a step into the unknown. The front of the 600 is very similar to the Motocoupé both mechanically and in appearance. A lengthened, wider chassis carries rear wheels of almost normal spacing (the difference between front and rear tracks is only about 2½ in.), and behind the divided rear axle is an o.h.v. opposed twin-cylinder engine of 585 c.c., derived from the well-known motorcycle unit and detuned.

The car is light—a dry weight of 10.1 cwt. is quoted—and from its behaviour on the road obviously rigid. Both virtues must be largely due to the mounting of a simple pressed-steel body upon an equally simple welded box-section frame. Of five cross-members, the three most important transfer weight to the all-independent suspension whilst one additionally carries the forward mounting of the engine and transmission.

That the power unit should be fitted at the rear is no more than orthodox in a modern German small car driven through its rear wheels. Inherently balanced and naturally suited to the compact space available, the flat twin engine itself is suspended behind the chassis frame, the major part of the weight being carried by a beam which is supported at its two ends by rubber bushes in tubular extensions of the frame. Clutch, gearbox and final drive are integral with the engine in normal rear-engine fashion, with the end of the gearbox joined by rubber in shear to a chassis cross-member.

Much development work has been done on the engine to adapt it to its new purpose, the main visible differences being in the ducted fan cooling and carburetter and ignition systems. A centrifugal fan mounted directly upon the rear end of the crankshaft draws air from the engine compartment and blows it forward over the cylinders, whence it can if required be sent under the rear seats for interior heating.

With a weight of half a ton to be propelled by two cylinders, flexibility is naturally of the first importance, and the 28 b.h.p. of the motorcycle engine has been reduced to 19.5 b.h.p. at 4,000 r.p.m., with a flat torque curve peaking at 29 lb. ft. and 2,500 r.p.m. Twin slide-type carburetters are replaced by a single central carburetter of butterfly type on top of the crankcase. To offset possible condensation of mixture in the comparatively long induction pipes a form of hot-spot is arranged, by bleeding air from a point in the cooling ducts adjacent to the exhaust port of each cylinder, and directing it to a jacket which surrounds the beginning of the inlet pipes. Typical of the clear thinking which is evident in the design is the fact that only a few inches of flexible pipe are necessary to feed the carburetter from a combined filter and reserve tap, screwed into the five-gallon fuel tank which fills the top of the engine compartment. A double ignition coil and low-tension distributor are used, for which current is supplied by a battery and silent Dynastart.

A single-dry-plate clutch, operated by enclosed cable, transmits power to an all-indirect, all-synchromesh gearbox (in contrast with the sliding-dog type of

B.M.W.-ISETTA 600

A wrap-around bumper distinguishes the front. Extra luggage space can be provided by folding down the rear seat squab, or removing the seat altogether.

the Isetta Motocoupé) which, in combination with the final drive, has overall ratios of 4.60, 6.90, 10.52 and 19.23.

Unusually, amongst its native competitors, the B.M.W. has no form of swing axle in its suspension, so that each drive shaft must be fitted with two rubber universal joints. The rear suspension is of pure trailing arm variety with fabricated arms pivoted on a chassis cross-member through widely spaced rubber bushes. The greater part of the load is carried by coil springs surrounding telescopic shock absorbers, but in addition the rear suspension is given a progressive effect by large rubber buffers which are compressed quite early in the 6¼ in. of spring travel. Front suspension follows exactly the same lines as that of the smaller model, using mainly common components in its coil-spring Dubonnet layout. As the complete spring and damper units pivot with the steering a divided track rod is unnecessary, and the simple steering linkage is not affected by variations in road surface. A refinement which further improves the front-opening door arrangement is a steering column with double universal joints, at top and bottom, so that with the door open the column is right out of the driver's way.

The inside of the car is practical and by no means austere. Examples of the intelligent use of space are combined arm rests and side pockets, and the spare wheel mounting under the lining of the front door. For comfort the front wheel arches are covered with plastic material, while for convenience there is a separate switch for flashing the headlamps as a warning.

A short trial of the 600 in minor roads around the Starnberger See, near Munich, revealed that four full-sized adults can indeed be carried comfortably. The 44-in. front seat can be moved back by a pair of wing-nut clamps to accommodate a six-footer (or its rake adjusted by two wooden blocks), and even in this extreme rearward position leaves just enough room for knees behind it. Luggage capacity with four seats in use is restricted to a space about 3 ft. × 9 in. × 1 ft. over the transmission, but it may be considerably increased by folding down the squab of the rear seat, or the hold made really cavernous by removing the rear seat altogether through the single, right-hand door.

With a load of passengers the riding was extraordinarily comfortable. There was none of the accustomed "rear-engine" feel, but rather a slight understeer, and sliding provoked on a wet road was quite stable and controlled. With a full load the car would start happily in second gear, and in fact left an over-riding impression that 600 c.c. of well-balanced and well-insulated air-cooled engine can lift a four-seat motorcar right out of the under-powered "miniature" class. There is reason to suppose that the manufacturers' claims of 62 m.p.h. and 47-63 m.p.g. are not greatly exaggerated. The new model will be produced concurrently with the 250 c.c. and 300 c.c. Motocoupé, of which some 75,000 have already been built.

BMW-ISETTA 600

Engine Dimensions	
Cylinders	2
Bore	74 mm.
Stroke	68 mm.
Cubic capacity ..	585 c.c.
Piston area	13.35 sq. in.
Valves	push-rod o.h.v.
Compression ratio ..	6.5
Engine Performance	
Max. power	19.5 b.h.p.
at	4,000 r.p.m.
Max. b.m.e.p.	120 lb./sq. in.
at	2,500 r.p.m.
B.H.P. per sq. in. piston area	1.46
Piston speed at max. power	1,790 f.p.m.
Engine Details	
Carburetter	Downdraught
Fuel pump	Gravity feed
,, capacity	5 gal.
Cooling system ..	Air, ducted fan
Electrical system ..	12v., 130w. Dynastart
Battery capacity ..	31 amp. hr.
Transmission	
Clutch	single dry plate
Gear ratios : Top (s/m)	4.60
,, 3rd (s/m)	6.90
,, 2nd (s/m)	10.53
,, 1st (s/m)	19.25
,, Rev. ..	18.75
Prop. shaft	Nil, rear engine
Final drive	differential and open drive shafts

Chassis Details	
Brakes	Hydraulic
Suspension : Front ..	Coil-spring Dubonnet
,, Rear ..	Coil springs and trailing arms
Shock absorbers: Front	Telescopic
,, Rear	Telescopic
Wheel type	Pressed steel disc
Tyre size	5.20-10
Steering gear	Worm
Dimensions	
Wheelbase	5 ft. 7 in.
Track : Front	4 ft. 0 in.
,, Rear	3 ft. 9½ in.
Overall length ..	9 ft. 6¼ in.
,, width	4 ft. 7 in.
,, height ..	4 ft. 6 in.
Turning circle	Approx. 26 ft.
Dry weight	10.1 cwt.
Performance Factors (At dry weight)	
Piston area, sq. in. per ton	25.4
Top gear m.p.h. per 1,000 r.p.m.	12.15
Top gear m.p.h. per 1,000 ft./min. piston speed	23.2
Litres per ton-mile ..	2,860

Although having a wheelbase of only 5ft 7in, the B.M.W. 600 can carry four adults in comfort, with some luggage. Side and indicator lights are mounted in the upswept bumper. There is a single side door for the rear passengers

FOUR-SEATER FROM WESTERN GERMANY WITH REAR AIR-COOLED TWIN-CYLINDER ENGINE

B.M.W. 600

NEW CARS DESCRIBED

THE range of German small cars has been increased by a new and interesting design. Bayerische Motoren Werke were known to be developing a novel small car, powered by their 600 c.c. flat-twin motor-cycle engine. Now, barely a month before the Frankfurt Motor Show, details of the car are revealed.

It is a full four-seater, with a front door giving access to the front seats, and a side door through which the rear seats are reached. A strong family relationship between the B.M.W. 600 and the Isetta Motocoupé is apparent. This external likeness is most pronounced when the 600 is seen straight from the front, for it is then difficult to distinguish it from the Isetta. From the back, the impression is of a well-proportioned and well-rounded, Multipla-like tail with a rear window of adequate width.

In Germany the basic price is 3,890 DM, or £330, which lies between the basic prices of the Fiat 500 and 600 models in Italy. Assembly and partial manufacture in Britain is planned for spring of next year; it will sell at £319 basic; £479 17s including U.K. purchase tax.

B.M.W. are fortunate in having for their new model an adapted and de-rated version of the horizontally opposed, twin-cylinder 585 c.c. motor-cycle unit famed for its smoothness, quietness and reliability. Inclined valves in light alloy heads are operated by push-rods from a central camshaft, driven by helical gears.

The engine is placed at the extreme rear, and cooling is assisted by a centrifugal fan, driven at crankshaft speed, which supplies air through ducts to the cylinder barrels and heads. A thermostatically controlled shutter at each air exit regulates the engine temperature. Starter and dynamo are neatly combined as a unit with the cooling fan; ignition is by coil for each plug, with a low tension distributor.

A single horizontal carburettor mounted above the light alloy crankcase supplies both cylinders. To minimize fuel condensation in the long induction pipes, the central portion is jacketed and fed with heated air from the cylinder ducts. Both exhaust pipes are led to a silencer below the right-hand cylinder. Air bled from the cooling duct passes through a heater coil in the silencer to supply warmed air to the interior of the car.

Clutch and gear box, with the final drive unit between them, form a single unit with the engine. The all-indirect, four speed and reverse gear box has synchro-

Unconventional entrance by the counterbalanced front door. Gear lever and hand brake are central; large side pockets are padded as arm-rests

Cooling air enters the engine compartment through louvres above the rear wheels

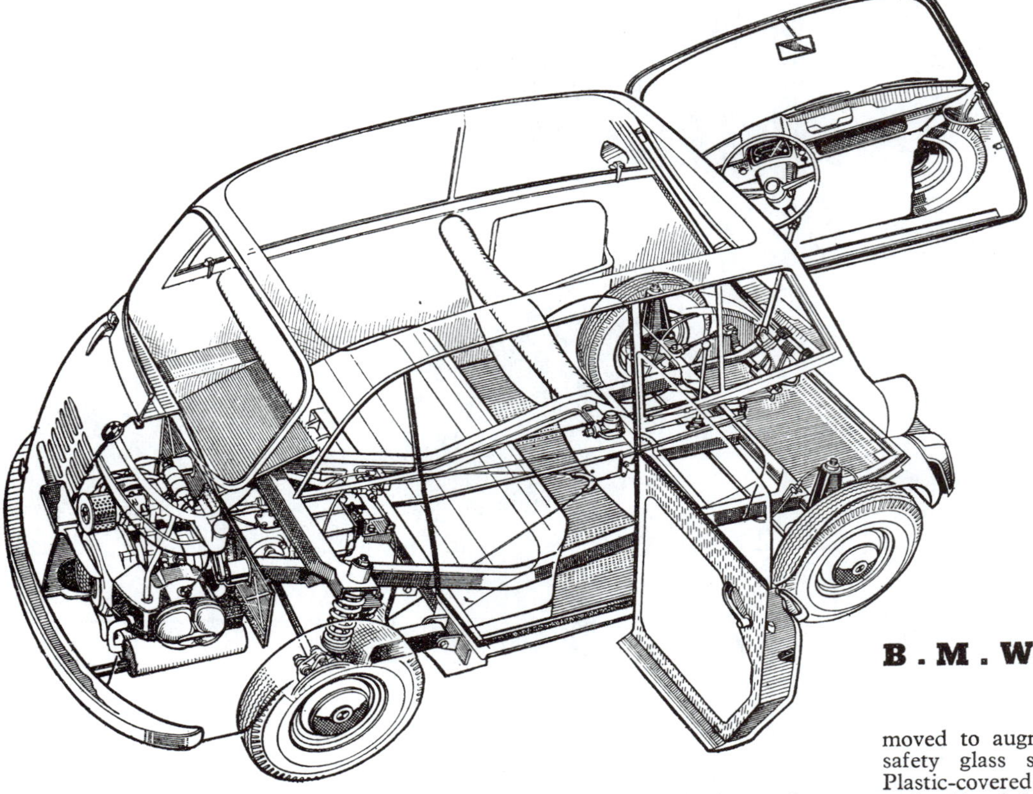

The entire length of the simple box-section chassis is available for passenger and luggage accommodation. Engine and fuel tank are behind the rear wheels, and the spare wheel is in the front door

B.M.W. 600...

mesh for each forward gear and the selectors are linked by a rod beneath the floor to a central, rigid, floor-mounted lever.

The simple chassis has box-section side members, swept upwards towards the rear to clear the universally jointed drive shafts; at the extremity of each member the power unit is supported on a large diameter rubber bush. At the front of the gear box is a third mounting consisting of a rubber block in vertical shear. This is carried on a substantial box-section cross member, to which are pivoted, in rubber bushes, the rear suspension trailing arms. Coil springs with concentric telescopic dampers, mounted on the rear cross member, are inclined forwards to give the desirable rising rate with upward wheel movement.

At the front the chassis is similar to that of the Isetta, though the track is 4ft —1in wider. Steering box and pedals are supported on a tubular structure ahead of the front wheel centres, driver and front passenger being placed well forward. Forward of the front seat, a tubular cross member carries at its extremities the steering swivels for the Dubonnet-type suspension assemblies. Each consists of a light alloy housing and two forward-facing, parallel-acting radius arms which support the stub axle, the upper arm acting through a bell crank on a coil spring mounted above it. Steering is by fore and aft rod to the left-hand assembly, and a track rod links the housings.

Built from steel pressings, the body has one door at the front of the car; side doors would necessitate stepping over the front wheel arches to enter the front seat. Facia and steering wheel are mounted on the door and the steering column, of course, is universally jointed and splined.

A side door gives access to the rear seat, which is well forward of the wheel arches. There is ample leg, knee and head room for two normal-sized adults in front and rear compartments and the plastic-upholstered seats offer a very adequate standard of comfort and support. Releasing two thumb screws allows the front seat to be adjusted; in its rearmost position there is still sufficient knee room for rear passengers.

Between the rear squab and the engine bay is a carpeted space which will hold a large suitcase and other soft luggage, and if need be the rear seat may be removed to augment this space. Sliding safety glass side windows are fitted. Plastic-covered padding is attached to the underside of the roof panel.

Levers beneath the steering wheel operate, on the right, the flashing indicators and horn (by lifting) and, on the left, head light dipping and flashing (by lifting). Choke and heater controls are on the wheel arch; the rest are on the facia.

A brief drive in the B.M.W. 600 showed that high standards of quietness, smoothness and tractability have been achieved. Synchromesh on all gears was efficient with normal use, and the changes were easy and pleasant to make. Visibility and manœuvrability were exceptional and the car was stable in the wet.

Power output is given as 19.5 (nett) b.h.p. at 4,000 r.p.m. The motorcycle version of the engine produces 28 b.h.p. in standard and 35 b.h.p. in sports versions. From this it will be seen that the de-tuned engine is working well within its capabilities and should have exceptional reliability and a long life. The makers claim speeds of 28 m.p.h. in second gear, 43 m.p.h. in third and 63 m.p.h. in top, that the car can be cruised at its maximum without harm to the engine, and that fuel consumption is in the region of 45-55 m.p.g.

SPECIFICATION

ENGINE

No. of cylinders ...	2 horizontally opposed
Bore and stroke ...	74 × 68 mm (2.91 × 2.68in)
Displacement ...	585 c.c. (35.7 cu in)
Valve position ...	Overhead, push rods and rockers
Compression ratio ...	6.5 to 1
Max b.h.p.	19.5 at 4,000 r.p.m.
Max b.m.e.p.	122 lb sq in at 2,500 r.p.m.
Max torque	28.9 lb ft at 2,500 r.p.m.
Fuel tank capacity ...	5.1 Imperial gallons (23 litres)
Fuel Pump	Gravity feed
Cooling system ...	Centrifugal fan
Battery	12 volt, 31 amp hr

TRANSMISSION

Clutch	Single dry plate
Gear box	Four speeds, all synchromesh, central lever on floor
Overall ratios ...	Top, 4.59; 3rd, 6.89; 2nd 10.53; 1st, 19.22; rev, 18.73
Final drive	Ratio 5.43 to 1

CHASSIS

Brakes	Hydraulic, leading and trailing shoes front and rear, mechanical handbrake on rear wheels.
Suspension: front ...	Independent, Dubonnet type with coil springs
rear ...	Independent, single trailing arms and coil springs
Dampers	Hydraulic telescopic
Wheels	Pressed steel
Tyre size	5.20—10in
Steering	Worm and nut

DIMENSIONS

Wheelbase	5ft 7in (170 cm)
Track	Front, 4ft (122 cm)
	Rear, 3ft 9½in (116 cm)
Overall length ...	9ft 6in (290 cm)
Overall width ...	4ft 7in (140 cm)
Overall height ...	4ft 6in (137.5 cm)
Turning circle ...	26ft 3in (8 m)
Dry weight	10.1 cwt (1,135 lb) (515 kg)

World's cheapest car to buy and run!

£359.19.6 inc. P.T.

or £120 down on easy terms

(Also Pickup van at £299.18.0 inc. P.T.)

If you're fed up with inflation, shocked at the cost of living, sick of counting every penny—here's news to set your heart dancing: a *real* car with 4-wheel safety and a proper chassis for £84 less than the next cheapest! British-made at that!

70 M.P.G.

But Isetta economy goes on and on: 70 miles to the gallon of petrol at 50 m.p.h.—tubeless tyres that last up to 40,000 miles—only one plug and six greasing points—3-pint oil sump capacity—a 4-stroke air-cooled engine that's trouble-free famous . . . why, it makes motoring almost as cheap as breathing.

FAST AND RUGGED

Add that the Isetta is *fast and rugged* (on official trial it covered the 2,350 miles from Sicily to Sweden in 62 hours, averaging 37.9 m.p.h., stops included!); that it's beautifully finished; that it seats two adults and a child in comfort; and that it beats the parking problem all ends up; and you'll know just why there's already a waiting list.

BRITISH-MADE

BMW

Isetta

'Best of all the economy cars'
Daily Herald

'It's a world-beater'
Daily Mail

'And the corners and bends? Taken steady and stable as a rock.'
Evening Standard

You have an Isetta dealer near you. Book a demonstration NOW!

Isetta of Great Britain Limited, Brighton 1

CONFESSIONS OF AN

EGG OWNER

I'M JUST an ordinary man, exceptional in no way: neither famous nor talented nor rich. Under usual circumstances, I'm not the least bit adventurous. Conservative in tastes, I look for the orthodox in life. And when it comes to automobiles, I like something not too different from the machines my neighbors own. Yet I have never been consciously guilty of trying to keep up with the Joneses. I'm just an ordinary man.

Or rather, I was, B.E. – Before Egg. The Egg in my case is an Isetta – a BMW Motocoupe, what they call in Europe a scootermobile. I own one, and that's my point. People of my ilk just don't. It's out of character.

It all started with our need for a second car. Not just any second car, but one that was economical to buy and run, easy to maneuver through heavy traffic, and suitable for my wife. Something she could use for taking the kids to school and doing the marketing. The need for this vehicle arose unexpectedly. My wife – otherwise a fine woman – simply announced one evening that she had made her last run for the 8:11. We live out in the country – a part of Connecticut that's called exurbia by people who live elsewhere – and I commute to New York: so it was either buy a second car or hire a chauffeur. That was that. Trouble was, I couldn't see buying another monstrous sedan – we already half-owned a Ford. I'll be frank: the problem was a financial one.

My teen-age son suggested a solution. What we needed, he said, was one of those cool little European bubble cars – a cross between a motorcycle and a midget car. A scootermobile. He showed me the ads in the New York papers. "Isetta 300 – World's Most Talked About Small Car – World's Cheapest Car to Buy and Run!" From the illustrations I could see that it was something you might talk about. But would it fly? Or rather, would it sustain life in crowded city traffic? In my jaundiced eyes, it had the makings of the most deadly suicide weapon since the Brooklyn Bridge. You certainly couldn't trust your wife and kids to it – it was just too small. In Connecticut we have lots of Buicks. And Packards. And Oldsmobiles. My God, this contraption would be run down by Volkwagens! I said no.

Next, my son brought me an ad from some British motoring journal: "If you're fed up with inflation, shocked at the cost of living, sick of counting every penny, then here's news to set your heart dancing: a *real* car with four-wheel safety and a proper chassis for £84 less than the next cheapest!" The ad – and my son – went on to expound the Isetta's many virtues: 70 miles to the Imperial gallon at 50 miles per hour – tubeless tires good for 40,000 miles – four wheels – room for three (small) adults – unobstructed four-way vision – direct access to and from the pavement through a single, front-opening door – and a trouble-free, air-cooled one-cylinder engine.

"*One* cylinder!" I exclaimed incredulously. "In the year 1957 I am to return to a motor vehicle having but *one* cylinder? Never!"

The economy angle did appeal to me – the Ford was getting about thirteen to the gallon, not to mention oil. But really, a one-cylinder, *one-door* coupe! Until then two doors had always been my idea of utter poverty.

From out of nowhere came a salesman – again the work of my son, who by now had become the scootermobile's greatest living exponent without having ever seen one in the flesh. This first salesman was followed by a second, then a third. Scootermobiles had obviously staged a successful invasion of our shores. BMW, Messerschmitt, Heinkel, Goggomobil. I now knew what had become of the German armament industry.

The Isetta salesman was most effective – he was the only one offering parts, service and immediate delivery. On top of that, he arrived in a demonstrator – a machine he called the USA Model 300 De Luxe. I could hardly believe my eyes. At first it looked like an Easter egg on wheels – whence the

Continued on next page

nickname Egg. I walked around it once, but the change in angle didn't help. There was something of Professor Beebe's bathysphere about it. Certainly it looked better qualified for excursions on the ocean floor than for shopping trips in Connecticut. Obviously it was related more to airplanes or motorscooters than to automobiles. The construction was typically aircraft: a tubular steel frame covered with sheet metal, glass and canvass. Such a concoction had no place in my life. I could see little method behind such Teutonic madness.

The salesman was smart. He said very little, but what he did say hit home. "This is the perfect answer to second-car needs. It's fun to drive, safe, economical, takes half the parking space of a conventional car, does sixty with ease, and you'll get forty miles to the gallon. Also, anybody — but *anybody* — can afford it." To hear him tell it, the Egg was kind of a poor volks' Volkswagen — costing some $800 less than its formidable rival from Wolfsburg. I asked what the exact delivered price was (the ads had said $998, but we all know what that means), and the salesman spoke right up: $1100 with radio and a few extras. I said no. In addition, I tossed in a little philosophy: while the Egg was certainly not a conventional vehicle, *I* was a conventional man. This had less than no effect. The salesman countered with the "take her for the weekend — no obligation to buy" ruse. He didn't even take us for a demonstration ride — just presented us with the owner's manual (in German) and left via taxi.

There it sat, cluttering up the driveway. What could I do? I decided to become better acquainted with the little devil.

The Old World craftsmanship was immediately evident — no file marks, no paint drippings around the edges, no slovenly metal work. I had to admit the little thing was beautifully made. But the thought of a head-on collision with even a bicycle was paralyzing.

Getting in was no problem — the whole front end opened wide and we just walked in, turned around and sat down. Me, my wife and the smallest of the three kids. I began by crushing my good hat, and a portion of my head, against the tubular cross-member over the driver's seat. My own fault, everybody said. The engine started up very quickly and had the sound of a power lawnmower, only deeper. It turned out to be directly under the seat, and it produced a full *thirteen* horsepower. The hand brake was located just under my left thigh. At least you couldn't forget to release it before moving off because you couldn't get at the clutch until you did. In fact, you could never forget it completely, even when it was off.

Brake, clutch and accelerator pedals were where you'd expect to find them, only smaller. The gear shift lever was on the left and seemed to grow out of the front wheel well. There were four forward speeds and reverse—each well marked and the whole thing just the opposite of any pattern you might already know. And all done with the left hand. A lot like BMW motorcycles, my son told me.

With much ado, we were off . . . with a neck-snapping lurch. I didn't quite have the "feel" of the clutch, and my feet were obviously too big. Once running, however, things went pretty well. I was immediately surprised at the sense of security and stability the Egg induced. There was nothing flimsy about it, despite its lightweight construction. Felt so good that I quickly lost control at the end of our driveway and entered the highway broadside-to. In sight of family and friends. Unfazed, I shifted up to second, making mild passes at fourth and reverse on the way, and settled down to a cruising speed of about seventeen miles an hour.

You have to drive an Egg to believe it. Words alone will not do. It was a lot like riding in a portable fishbowl. Steering and handling turned out to be lots of fun. Finding true third and then even fourth, I got her up to fifty, then almost to sixty on a slight downhill grade. For several brief moments I felt master of my fate. But it didn't last. Heading for home on a secondary road, I found that rough stretches of pavement produced a profound pitching and yawing, and it was necessary to slow to twenty-five in order to survive. The wheels were very small—4.80-10's—and the Egg handled like a large baby-buggy overloaded with groceries.

The salesman had said much about the Egg's "lively" performance. What is "lively" to one man is snail-pacing to another. Fortunately, I'm a snail-pacing kind of guy. Whatever may be said of actual performance, it *sounded* like it was really moving. And up to about thirty, I guess it really was. The engine seemed to be quite slow running, and you knew it had only one cylinder. It was enough to keep you from being run over, but not much else.

When the salesman returned Monday evening, I was convinced: we didn't want an Egg. The salesman was glib and, to the rest of the family, convincing. I ended up signing on the dotted line—but for only one very good reason. The Egg was so scuffed up from the weekend's trial runs that I was ashamed to return it. No obligation to buy—indeed!

We owned an Egg—perhaps the first in all Connecticut. And for the first week everything went well. While the novelty lasted, my wife and kids couldn't see enough of our shiny red scootermobile. Each evening I'd come home to find the driveway crowded with teen-agers taking turns at the wheel.

But by the weekend, things had changed. The Egg was to be *mine*. Wife and kids wanted the Ford back!

"What do you mean, *I've* got to use it? We bought it for you and the kids, remember? For going to school and doing the shopping."

My wife's reply was devastating in its lack of logic. She and the girls found the Egg *embarrassing*. They felt funny, stared at, self-conscious, too out of the ordinary! The girls simply refused to be seen in it: "Mom, if you're driving the Egg, don't bother to pick us up. We'll hitch a ride home." My wife's bridge club friends concurred: the Egg was an eyesore. What's more, my mother-in-law forbade my wife to drive it on Main Street — it brought shame on the entire family. That was the killing blow.

The Egg was mine, for better or worse—by popular demand. My keys to the Ford were confiscated. It was either the Egg or nothing. Reluctantly, I set out to make friends with it. After all, it stood between me and the train station.

The Egg's attention-getting properties were its most serious drawback. It wasn't just transportation, every ride was an event. Suddenly I was an object of public ridicule and merry-making simply because I'd had the ill fortune to inherit a beastly little freak from across the seas. People stared. Not just the children, but the pointing, gesturing, gesticulating parents as well! Some shouted words of encouragement. Others treated me like the brave, demented fool I must have been.

With no warning, I found myself traveling thirty miles per day by Egg—fifteen to the station, fifteen back. At first even the simplest maneuver was a challenge. I'd have been as much at home in the Nautilus. I couldn't tell which gear was most appropriate at which speed. I'd try second, decide the engine noise level was too high, find third, discover that my speed had fallen off to twenty miles an hour, and return to second and those screaming gears again. First was hard to find when you were at a standstill and impossible to engage when you were moving. Second was hard to find when moving, and you didn't want it from a standstill. Neutral seemed to be most effective of all.

Arrivals and departures were a nightmare. On the open road you could always ignore the idle threats and exortations of passers-by. But at the station, where you actually had to get out and lock up or unlock and get in, you were vulnerable. I prepared some stock replies. "Yes, *I* like it." Or, "She *is* a pretty little thing, isn't she?" Or, "Gets me where I'm going, and that's the main thing, now isn't it?" Or, "Wouldn't be without one — actually, this is my *second*." It helps if you're a born liar.

Performance—or the lack of it—became a matter of constant embarrassment. English racing bikes, some of them with mere women aboard, made a habit of whipping by me on a particularly bad hill near home. I ended up going miles out of my way to avoid the field of combat.

Radio reception was poor, so I had one

of those long, whip-type antennas rigged up in back. This was a mistake. "Hey, fellah, what are yuh? Radio controlled or somethin'?" Or, "Watcha got there, Mister — one o' them intercontinental ballistics missiles!" I took to working late at the office in order to return home after dark.

Parking was no problem—the Egg being only 7½ feet long, 4½ feet wide and (in case you wanted to park it *under* something) 4 feet and 4½ inches high. However, at the station it was important to park well back in the parking space and not with the front bumper against a retaining rail, as others did. Otherwise a few unsuspecting souls would turn confidently into an "empty" space, only to find at the last tire-screaming moment—the Egg.

And there was the matter of the surreptitiously planted notes. For the first month I'd leave the train, walk to the Egg, and find several "good-natured" messages under the wiper blades. Some were scribbled on business cards, others on the backs of old envelopes, others on wives' shopping lists. Some were even typed in advance — probably dictated to some highly amused secretary. "Got to hand it to you, Mac—you got guts!" Or, "What's it going to be when it grows up?" Some were frankly insulting: "Your car (?) here plus $3000 cash and I can put you in a brand-new Chevy Six Two-Door. Act now—the market for your bug can't last!"

Yes, the Egg seemed to induce high-spirited playfulness in people of all ages. My wife and I left the movies one evening to find our poor little Egg beached high and dry on the sidewalk, carefully wedged between a store front and conventional car parked at the curb. College boys, the theater manager said. After all, the Egg weighed less than 800 pounds—a small load for four or five healthy young men. We—the manager, four or five healthy young men, and I—moved her back into the street; and we—my wife and I—learned to watch TV.

Unable to beat the Egg — no pun intended—I joined it. I stopped fighting the situation. The change was psychological. I bought one of those natty caps with the belt in back and found it gave me more self-confidence. (If you're going to act like a fool, might as well look like one, too.) Soon I didn't mind being stared at—in fact, I encouraged it by smiling openly at other motorists. I took to doing extra shopping on weekends just to be seen around town more. I acted as though there was nothing to be ashamed of.

I learned a few tricks of the trade and applied them regularly with good effect. Parking nose-to-the-curb between two conventional cars, for instance. Preferably on Main Street. This was a real crowd pleaser, especially on Saturday mornings in front of the hardware store.

When fortunate enough to have a passenger new to the Egg — and believe me, most passengers were — I went through another good piece of stage business. Behind the seat was a control not found in "ordinary" cars — the reserve gas supply. (There was no gas gauge, and the main tank held only three gallons — enough for maybe 180 miles. The reserve tank held another eight-tenths of a gallon.) At the first noticeable miss of the engine, I'd say to my passenger, "Oh, oh — time to switch over to the reserve," reach back—without looking, mind you—and flip the lever. All very professionally done. *Wonderful* effect.

Then the inter-office memo came from the Boss. He and I had never been close—matter of fact, we'd only met once, the day I was hired. Now he wanted us to come to supper. Scrawled at the bottom of his note were these ominous words: "And bring along that *thing* of yours. Want to see just what you've been up to."

We drove the six miles to the Boss's place in a state of panic. Obviously we'd committed a breach of etiquette and social good sense by acquiring our Egg. Now we were going to pay for it. We skittered to a noisy stop at the Boss's front door, prepared for disaster.

Surprise, surprise—the dinner was a huge success. I made one bad mistake—giving the Boss his demonstration ride *before* supper and *after* two drinks. He couldn't get enough of it. One ride around the neighborhood and he asked to solo. After a couple of ear-shattering gear changes (he didn't understand the clutch either) he had the hang of it. After fifteen minutes of sprints around his long circular driveway, he had mastered what owners call "the Isetta slide" and gone on to bigger and better things. I've never before seen a man so quickly and willingly seduced by a piece of machinery.

Three days later and mine was not the only Egg in our part of Connecticut.

Today, if I may say so, I am a new man. Ordinary, except for my Egg. Where once my Egg and I were strictly an odd-ball combination, today we are a much-respected, even beloved landmark in the community.

Today, in my quite way, I am something of a civic leader. Thrice I have been honored at luncheons proclaiming my virtues as a pioneer, a pace-setter, a man who — despite initial opposition from certain bigots and narrow-minded reactionaries—led the way to a more civilized, sensible mode of transportation. I have been elected to the Chamber of Commerce, the Civic Betterment League—and yes, I am president of the local Isetta Owner's Club. Truly, I have prospered in business, in my community, at home.

I blame it all on the Egg.

People no longer laugh at my Egg. Most sensible machine, they now say. Shows lots of common sense — and a good dash of Yankee ingenuity. There's just no sense in driving some gigantic sedan back and forth to the station. Yes, sir, most sensible machine.

I am happy to report that we have even had our first exclusive Egg Owner accident — a breath-taking collision right in the station parking lot. Two broken headlights, a dented body shell and some scratched paint. The two participants were seen to embrace moments after the crash. That's what Egg-owning can do to you . . . I mean *for* you.

Brown Meggs

ISETTA Italy

ISO S.P.A., MILANO

Among the world's smallest and most originally designed automobiles is the tiny, front entrance, two-seater Isetta. This fabric-topped car offers remarkably low fuel consumption, traveling more than 65 miles to the U. S. gallon of gasoline, at an average speed of 40 mph.

The chassis consists of tubular components that extend beyond the front wheels, and the car's overall body length of only 88.58 inches enables the Isetta to be parked practically anywhere.

It features an independent front suspension with coil springs, while the rear suspension is a rigid axle on leaf springs.

The powerplant is an off-center, rear-mounted, inline coil ignition engine of 14.4 cubic inches. This two-cylinder, two-cycle engine develops 9.5 horsepower at 4500 rpm, and is suspended on three double pairs of rubber anti-vibration pads.

The roomy seat comes equipped with an adjustable back rest, and both of them can be readily removed so as to be used for picnics or camping. The Isetta, with its four wheels, four-speed transmission, hydraulic brakes, comfortable seating capacity for two, and ample luggage room behind the front seat, meets most basic requirements of a passenger automobile. ■

8 reasons why the

Isetta is in such demand

(*Basic model*)

It is the world's cheapest car to buy and run.

It costs pounds less than the next cheapest.

It does 70 miles to the gallon at 50 m.p.h.

It has tubeless tyres lasting up to 40,000 miles.

It has a robust, smooth-running 4-stroke air-cooled engine.

It seats 2 adults and a child in comfort.

With its front-opening door, it is the easiest car in the world to park (nose to kerb).

It gives 4-wheel motor-car safety with motor-cycle economy.

(*Prices include Purchase Tax*)

Standard Saloon	£383 . 19 . 6
'Plus' Model (Front Bumper Horns, Rear Bumper with Overriders, Side Mirror,Pocket on Door, Grab Handle, and Hub Caps)	£399 . 19 . 6
'Plus' Model fitted with Smith's 'Selectroshift' 2-pedal only transmission	£424 . 19 . 0

(*'Plus' model*)

ASSEMBLED IN BRITAIN—

91 British firms supply material which represents 64% of the ex-works selling price of the vehicle.

The remainder of the parts are at present being imported from Western Germany.

ISETTA OF GREAT BRITAIN LIMITED, BRIGHTON 1

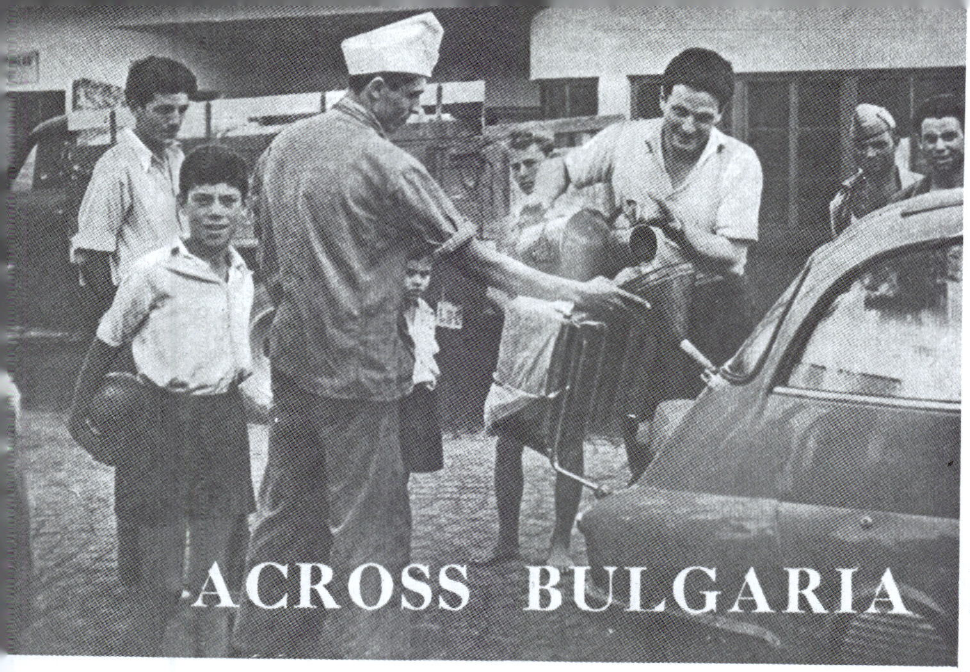

Refuelling at Bourgas with the "best" petrol—66 octane

Not what even the Bulgarians would call a first-class highway. . . .

ACROSS BULGARIA IN A YOGHOURT JAR . . .

ISETTA FOR TWO—AND A HOLIDAY JOURNEY TO THE SHORES OF THE BLACK SEA • By Paul Niedermann

EVERYONE knows that miniature cars are becoming more and more important on the European market. It seemed essential to test one, so my wife and I decided to use our holiday to try out a small car. We were determined not to sacrifice all our enjoyment for the sake of the test; we needed sun, sea and all the rest. Then, too, a proper trial should include driving over bad roads and on all sorts of fuel. There was only one solution—the Balkans.

As we already knew Yugoslavia and Greece, we decided to try our luck behind the Iron Curtain, and Albania was our first choice. After a ferocious, six months' siege of the legation attached to this apparently hermetically sealed "popular republic," we had to abandon the idea of Albania and turned to another country in the Eastern Bloc—Bulgaria. A kindly press attaché smoothed the way for us and visas were procured in record time.

We still had to find a little vehicle capable of standing up to shock treatment and bringing us back safe and sound. Hoepner, B.M.W.'s chief press officer, promised us an Isetta 300 for the chosen date. We thought the Isetta was eminently suitable for our destination because its shape is so exactly like a yoghourt jar—yoghourt being, of course, the staple food of Bulgaria, and held in great respect there.

I must admit that we had our doubts about the Isetta, which, after all, is meant for town transport and not built for cross-country driving. We arrived at the B.M.W. factory at Munich at 8 a.m.—we had warned their public relations office about the trip, asking for a well-prepared vehicle. We were rather dismayed to find that a German colleague had handed our model back the evening before, the distance meter registering 14,000 km—and 14,000 kilometres of road testing are as rigorous as twice that number of ordinary kilometres. As there was no time to spare, the Isetta got a change of oil and a wash; that was all. Half an hour later we were on the *autobahn* making for Austria.

I have not space here to describe our drive through Austria and Yugoslavia—enough to say that our Isetta leapt over the Grossglockner and climbed the 1-in-3.6 slopes of the Loibl pass up to the Austro-Yugoslav frontier.

We left the main highway at Nis, and from there on we were on our own, except for a few peasants on their donkeys. Soon we were stopped by a locked gate guarded by a soldier, bayonet fixed—the celebrated Iron Curtain. The gate was opened for us, but clanged to again as soon we had passed. This was a chilly welcome, but Bulgarian customs men made up for it; they were the friendliest we have met in all our travels. They asked how much money we carried, noted the amount down on a form, glanced at our carnet, and that was the end of the formalities. The officer in charge spoke a few words of German and asked us endless questions about the Isetta. Before we left he reassured us about the road surface ahead and, in fact, this was very good. Small cobbles alternated with tar and cement as far as Sofia—but on this stretch we did not meet a single car.

We arrived at the Bulgarian capital at midday on Saturday. This was a catastrophe—nobody works in Bulgaria on Saturday afternoon, and wherever we went we were surrounded by hordes of people. It was impossible to stop without being encircled by at least 100. Then we could not get petrol coupons, cash or anything else, and we were forced to stay in Sofia for the weekend.

After the destruction of the last war the centre of Sofia was

The Iron Curtain—along the lane behind the barrier is the Bulgarian frontier post, our first on the road to the East

Across Bulgaria in a Yoghourt Jar . . .

completely rebuilt, the new buildings, mostly housing and government offices, contrasting strongly with the houses outside the new ring and in the suburbs. There are, of course, many museums and churches. I must mention a little church on the outskirts of Sofia. A few kilometres from the capital rises Mount Vitocha, 2,000 metres high. On its flanks, only 4 km from Sofia, is the village of Boyana. Its church was built in three parts during the tenth, twelfth and nineteenth centuries. The beautifully kept mural paintings of the tenth and twelfth centuries are worth a visit by themselves. Mt. Vitocha itself is a perfect paradise for campers, the enchanting site abundant with streams. Many picturesque restaurants welcome hundreds of visitors each Sunday.

On Monday morning we went first to the Militia to give them a list of the places we planned to visit. This formality is insisted upon, but usually one gets authority to travel without difficulty. The official tourist organization—christened Balkantourist—gave us all the petrol coupons we asked for, which saved us a lot of money—about half the price of each gallon. As for oil, it is as well to carry a good supply because Bulgarian oil is rather poor and the price is astronomical.

Before leaving for the interior of the country we changed a good deal of our money into Bulgarian *leva*. The currency is another story—on the day we left Bulgaria the exchange rate of the *leva* rose by 40 per cent for tourists. Replete with money, petrol and oil, we set off towards the Black Sea.

The main highway of Bulgaria crosses the country from the Yugoslav frontier to Edirne, in Turkey. This is the only road with a well-kept surface. It is particularly useful for travellers to Turkey, because it is much shorter than the usual route via Macedonia and Greece. Crossing the fertile valley of the Maritza, the road goes through Plovdiv, the second most important town of the country.

Soon after Plovdiv we left the main road to wander along the Balkan by-ways. We saw many road works, but at present the road network in Bulgaria is still in its infancy. Late in the evening we reached Bourgas, a little port beside the Black Sea. It was full of summer visitors from Hungary, and it was only through the kindness of a hotel doorkeeper who spoke fluent French that we found a small room—it was too late to start looking for a camping site. There is no organized camping in Bulgaria, but one can choose one's own site practically anywhere.

Next day we had planned to go to Sozopol, a very old fishing village, founded by the Greeks, but the roads were so bad that after 17 km of indescribable jolting we returned to Bourgas.

Here we had the only breakdown of our journey. Strange noises from the engine warned us to check the oil level, and we were worried to find that the sump was practically empty—we had lost the breather and oil was escaping through a great hole. This incident—insignificant under ordinary circumstances—became high drama in the depths of the Balkans. In despair we bought a dozen stoppers at a chemist hoping to close the hole. The high pressure inside the sump literally blew out the first stoppers, but with the aid of a tube of glue we finally succeeded.

From then on oil seeped through every joint—but after all the engine must be kept lubricated. This greasing by "lost oil" cost us all of £25, because our reserve store lasted only as far as Sofia. Returning through Jugoslavia we found some good quality oil but at exorbitant prices—more than £1 a litre.

But to return to our travels: From Bourgas we went back up north to Nessebre. This enchanting little town, perched on what is practically an island in the Black Sea, was founded by Greek sailors at the beginning of the Christian era. These navigators must have been immensely rich, for nearly every family built its own church. If, today, Nessebre has not a single

Judging by this display of equipment the Bulgarian roads have a good chance of being made good

The bridge was made only for pedestrians and cyclists—and the road goes through the river (upper picture)

The car needed a good wash—and the bag of sweets which rewarded these young Yugoslavians was well earned

Map reproduced by courtesy of Hallwag, Bern.

There are no road maps of Bulgaria as we know them—in heavy lines here are shown main trunk roads linking with the capitals of neighbouring states

Left: Beside the beautiful Scutari Lake looking across at impregnable Albania. Right: Near the Albanian mountains we tried on the local headgear—it proved so useful in the sun and rain that we continued to use it until we returned to France

hotel, it makes up for it with no fewer than 42 churches in various stages of repair. It boasts also a magnificent, sandy beach.

The Black Sea is slightly saltier than the Mediterranean, so the water is buoyant. The climate is like that of the Côte d'Azur and the water is blue and incomparably limpid; it is called the Black Sea because of the violent storms which rage around its shores.

Because there were no hotels we had to camp. We found a promising spot in a spinney, and were given a great welcome by four Bulgarian families already installed there. As usual on our journeys, we had eaten Bulgarian food at a restaurant. Of course, if you ask for it, you can have "international" food with no particular character, but that is hardly the point of going abroad.

From Nessebre we went to Varna, the great port and fashionable beach of Bulgaria. During this lap of our journey we had our only trouble with the local police, our road apparently passing through a military zone without any indication that this was so. After a long discussion with the officer in charge, we had to return to Bourgas and take the old mountain road, which was worse than anything we had yet experienced.

Arriving at Varna late in the evening we were told that about 10 km outside the town Balkantourist had just built hotel buildings and restaurants, grouped in a park near a beach. Although it was then very late we finally found a beautiful room in one of these ultra-modern hotels—at 40 leva for the two of us.

Bulgaria is still at the very beginning of its tourist career, and the lack of experience is very evident. Still, the Balkantourist hotels at Varna are an example of what is being done. A Swiss hotel expert has been appointed to train and gather hotel and restaurant staff according to traditional methods—the poor man has a job!

Near Varna there are many museums and monasteries to visit, but our visas were running out, so we took the road back via Sofia, going north of the Balkan chain. The next stop was Trnovo, an ancient city on the banks of the river Jantra in the mountains. Trnovo played an important part in the history of Bulgaria, for it was the country's capital at the time of the first Bulgarian empire. It is certainly worth strolling down the old streets and visiting the national museum. The Balkantourist hotel has a splendid view of the Jantra and the surrounding mountains; prices are moderate and the people charming.

We left Trnovo to attack the highest pass of the region, the Chipka. It is only 1,400m high but the Bulgarians are very proud of it. At the top is a monument to the glory of the army of the Tsar of all the Russias, who freed the Bulgarians from five centuries of Turkish oppression in 1877.

In the mountains our engine showed signs of fatigue; we checked the main components, but nothing seemed wrong. After many tests we felt practically certain the trouble came from the oil filter, in which the filtering element was a strip of paper guaranteed for 10,000 km. It may well have lasted the distance under other circumstances, but on the Bulgarian roads. . . . Fortunately we had a spare, so we changed them every 1,000 km, cleaning them thoroughly each time.

Then we went down to Kazanlik, capital of the celebrated Rose Valley, the engine sounding healthy again, and continued on the same day towards Sofia. Two alternatives faced us, direct return through Rose Valley on a mediocre road, or a detour via Plovdiv on the road already described. We decided finally on the latter because of the state of the roads. In the end we did not benefit, because floods near Levskigrad forced us to deviate by 150 km over a positive obstacle racecourse. We patronized a completely new luxury hotel in Plovdiv at a reasonable cost of 40 leva—about £2 at that time.

The return to Sofia and the frontier was child's play, but we were not yet back in Paris; ahead of us lay Macedonia, Montenegro, the Dalmatian coast, northern Italy, the Italian riviera, the Côte d'Azur, the French alps, Grenoble, Lyon, Normandy and finally Paris. All that amounted to about 8,000 km—no mean distance under the circumstances. Back in Paris, we put the Isetta through its paces at Montlhéry. As it arrived from its marathon—with clogged filter, badly stretched throttle cable and without a breather—it reached 85 k.p.h. (53 m.p.h.).

Tyres, too, deserve special mention. We drove on Metzeler tubeless from the start, and for a month we simply forgot we had tyres, while the poor Bulgarians spend half their time mending punctures.

As an epilogue, here is some advice for those who may plan to go to Bulgaria: First, the Bulgarians are mostly charming people, both hospitable and helpful. As to exchange rates, it is difficult to give much information. When we left Germany the market value of the lev was strong compared with eastern currencies, but since 1 August, 1957, the rate has improved by 40 per cent for tourists and the situation must be good at present. Barter can be very interesting, some things being very rare; for example, watches, fountain pens, cigarette lighters, jewels and anything made of nylon are in great demand. There are good hotels in all the fairly important towns and you can safely camp anywhere, with one exception—we do not advise staying near a gipsy camp.

There is no such thing as a Bulgarian road map, so you must equip yourself as best you can this side of the Curtain. There are two kinds of fuel in Bulgaria, normal petrol and aviobenzine, but one is as bad as the other—that is to say between 60 and 66 octane—so it is advisable to adjust the engine accordingly. Most fairly important places have petrol pumps, but it is as well to keep an eye on working hours; if you remember that pump attendants usually work between 8 and 12 and again from 3 to 5 you will not go far wrong. The mere 13-litre tank of our Isetta was a positive nightmare, and it is best to carry a spare can. As many petrol coupons as you want can be obtained from Balkantourist, but that only applies in Sofia!

Finally, remember that from midday Saturday until Monday morning nobody works, not even a cashier or a pump attendant (except in Sofia). At the moment a great deal of roadbuilding is in progress. When it is finished the country will be properly open for tourists.

Thanks to this ten-litre reserve of petrol we were able to go from Tirnova to Plovdiv via the Chipka pass

NOW... EUROPE'S HIT CAR IS HERE

BMW ISETTA SPECIFICATIONS

Engine	BMW fan-cooled single cylinder four-stroke engine
Bore	72 mm (2.83 in.)
Stroke	73 mm (2.87 in.)
Capacity	298 c.c.
Compression	7.0 to 1
Power	13 hp
Valves	over-head valves, push rod operated
Lubrication	Circulating system with gear type pump
Starter	12 Volt starter
Electr. Equipment	12 Volt / 130 Watt Generator
Engine Position	Right hand side, transverse behind the seat
Clutch	Single plate dry clutch
Transmission	BMW four forward speeds and reverse gearbox attached to the engine
Final drive	Resilient mounted drive shaft and totally enclosed, fully adjustable chain drive in oil-bath
Chassis frame	Rigid tubular chassis frame
Wheel Suspension	Independent front wheel suspension, swinging arms, coil springs and hydraulic shock absorbers. Rear wheel suspension, quarter elliptic leaf springs and telescopic hydraulic shock absorbers.

Wheels	Split-rim disc wheels. Spare wheel with tire		
Tire Size	4.80 – 10		
Steering	Steering wheel, reduction steering gear, smallest turning circle approx. 24 feet		
Brakes	Four wheel hydraulic brakes. Brake diameter 7 in. Total brake lining area 49.9 sq. in.		
Dimensions:			
Tread, front	47.2 in.	Overall length	89.9 in.
Tread, rear	20.4 in.	Overall width	54.3 in.
Wheelbase	58 in.	Overall height (unladen)	52.6 in.
Weight:			
Kerb weight	approx. 770 lbs.		
Carrying capacity	507 lbs.		
Road Performance:			
Max. Speed	52.8 miles per hour		
Climbing ability	First gear 1 in 3		
Fuel consumption	approx. 62 miles / U.S. Gal.		

Modification to details and specification without reserve
Body design licensed by Iso s. p. a. Milan / Italy
Printed in Germany MC 122 e 100 10. 57

BMW Isetta

There's no easier way of getting in and out of a car, than through the wide swing-open front door of the BMW Isetta. Try it yourself – just step in standing up, turn, sit. As convenient as that . . . and no more crumpled clothes, mashed hats, ever.

Two adults and a child have room to spare in the ample front seat.

A rear carrier accommodates sport equipment and luggage.

Completely unobstructed 4-way vision is just one of the many outstanding Isetta features. And for fine weather, your Sun-Liner roof slides back for real convertible pleasure.

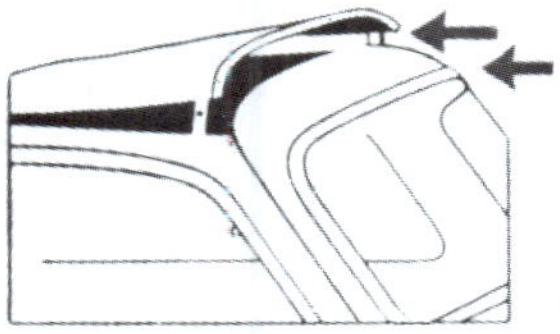

Adjustable air vent top permits comfort controlled circulation of air.

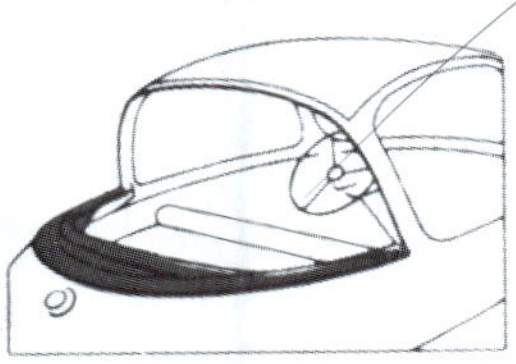

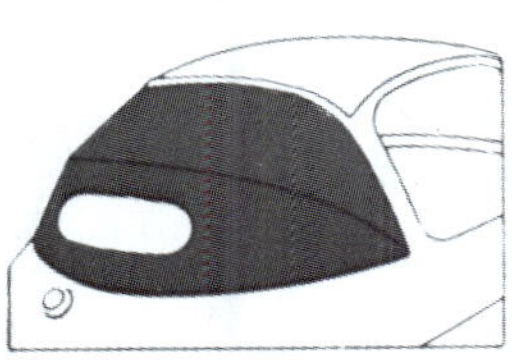

Standard Sun-Liner model with sunroof foldback canopy.
The BMW Isetta is also available with a convertible hood and an adjustable front door ventilation system.

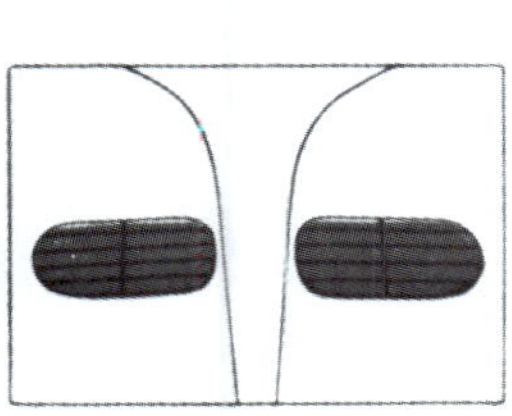

The BMW Isetta is the ideal second car – for shopping trips, commuting to the station, picking up the children after school. And the cost is so modest. The car almost pays for itself in gasoline, oil and maintenance savings alone. No matter where you live . . . there's an Isetta dealer near you . . . with a complete line of spare parts in stock. If you should ever need them, though Isetta owners overwhelmingly agree that they've never before had a car that stays road perfect for years – needs so little maintenance.

The 1955 Mille Miglia, – the tough 1000 mile Italian road race, proved the performance of the Isetta Motocoupe, which covered the distance at an average speed of almost 50 miles per hour. A performance which was acclaimed all over the world.

THE BMW *Isetta* IN DETAIL

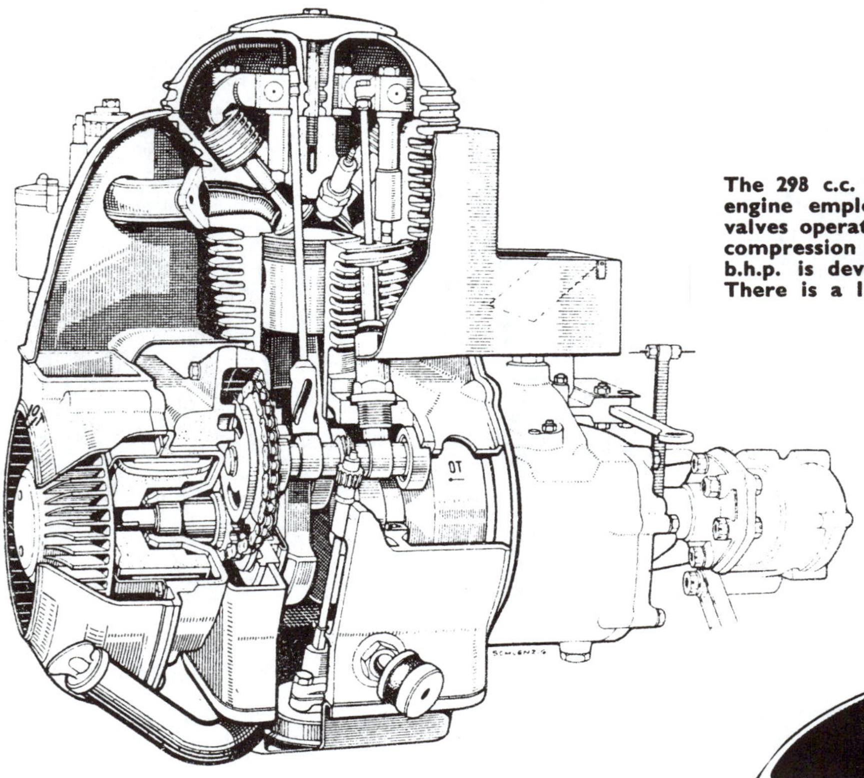

The 298 c.c. air-cooled four-stroke engine employs inclined overhead valves operated by push-rods. The compression ratio is 7·0 to 1 and 13 b.h.p. is developed at 5,200 r.p.m. There is a 12-volt dynamo-starter.

This drawing shows a left-hand drive version. The narrow rear track permits the use of a rigid rear axle. The gearbox provides four forward ratios. Final drive is through a transverse-mounted drive shaft and enclosed chain drive in oil bath.

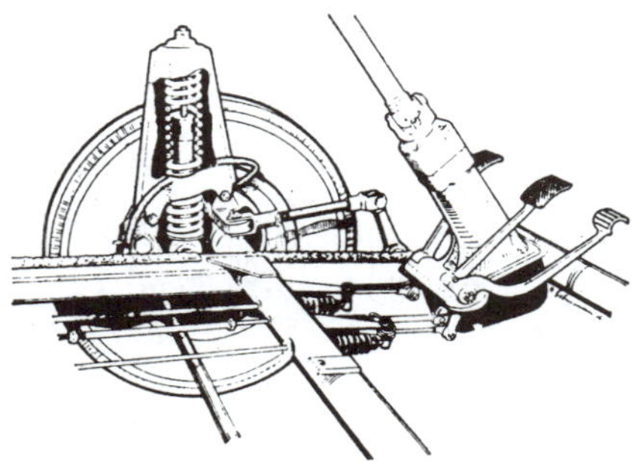

The arrangement of the Dubonnet-style independent front suspension is seen here — also the universal joint at the foot of the steering column.

A CAR ROAD TEST

The BMW Isetta 300 Motocoupe

IF there were a prize for originality of design the BMW Isetta would stand an excellent chance of winning it. Within the inescapable limits imposed by the scaling-down process, this miniature vehicle is surprisingly efficient. It is outstandingly economical, is a lively performer on crowded roads, provides an acceptable degree of riding comfort, has very good road-holding qualities — and, above all, it is practical and sound.

We have mentioned inevitable limitations due to ultra-small dimensions. In our view, the most important of these is the proximity of the engine to the driver. A single-cylinder air-cooled engine is never notable for either quietness or smoothness of operation, and the Isetta's power unit is positioned immediately behind the driver (in right-hand-drive versions; it is behind the passenger in left-hand-drive models).

It is fair to say that the amount of noise that reaches the driver's ears is as great as might be expected, but that there is less vibration than might be thought likely to result from such an arrangement. Travel is noticeably quieter on the passenger's side of the seat.

We found, however, that we became quite speedily accustomed to the noise, and, provided that we did not intersperse our Isetta with motoring in cars of more conventional size, we discovered ourselves becoming on occasion almost oblivious to it when cruising at 35–40 m.p.h.

With a wheelbase of only 5 ft. 1 in., one would expect a considerable amount of pitching motion on any road surface that was short of perfect. Additionally, with a difference of about 2 ft. 3 ins. between front and rear tracks (and a wheelbase little more than 1 ft. greater than the front track), one would anticipate encountering some odd sensations when cornering. It is a tremendous compliment to the designers that the Isetta is in fact remarkably free from pitching, and that it proves itself to be capable of virtually roll-free cornering at speeds appreciably above those at which many larger vehicles can safely be taken round curves and bends.

Good handling characteristics and well-positioned controls play a generous part in making the Isetta a safe little vehicle that is enjoyable to drive.

Provided that he does not sit bolt upright, a six-footer can make himself comfortable at the controls. The pedals are not positioned too close to each other, the steering wheel is set at a sensible angle, the levers controlling the gear-change and the hand-brake are excellently sited for ready operation by the right hand, and the minor controls (with the exception of the wind-screen-wiper switch) are within easy reach of the driver.

The steering is rather low-geared and suffers from some lost motion, but it is light and road shocks are not transmitted to the driver's hands. The amount of pressure that is required to operate the pedals seems, on first acquaintance, to be rather high for so small a vehicle, but once one's initial surprise at the discovery has worn off one ceases to be conscious of the matter.

Apart from the fact that a little easing of the clutch is sometimes called for in order to engage first gear when at rest, the gear lever works excellently; only a ham-fisted driver devoid of all sense of timing could fail to make smooth and quiet changes of ratio. Similarly, only an exceptional degree of ham-footedness could prevent a driver from making smooth starts: the clutch engages smoothly, and a heavy flywheel does much to counter the relatively long interval between power impulses that results from the employment of a single-cylinder four-stroke engine.

The flasher-type direction indicators are appreciated, for the sliding front-halves of the windows do not produce apertures through which hand signals are easily made. These turn-indicators are not self-cancelling, but there is the usual warning light (and clicking sound) when they are in operation.

With the exception of blind-spots caused by the stout frame of the single door at the extreme front of the vehicle, visibility in all directions is excellent. The fore-part of the canvas roof is readily folded back. We noted that the noise-level inside the little vehicle was definitely lower when the top was open.

Entry via the spring-counterbalanced front door is simple: one walks in, turns round, sits down on the edge of the seat and then settles back in a comfortable position. Exit is only a trifle less easy, it being necessary to edge oneself forward on the seat before standing up and stepping out.

A universal joint at the foot of the steering column enables the steering wheel to swing forward when the door is opened; the door is closed from within by pulling the steering wheel back into position, and its final closure is effected by an efficient — if somewhat stiff — handle. There is ample room on the seat for two adults, and one small child can quite easily be accommodated in addition if necessary. A small lever close to the interior handle of the door can be depressed to lock the door from the inside. This is of undoubted value when small children are being transported, for the door handle is on the passenger-side.

A key, different to that which operates the ignition switch and starter, locks the door from the outside. It is possible to park the Isetta with its windows open in hot weather without any unauthorised person being able to unlock the door.

A third key is used to free the engine-cover. Engine accessibility is poor, and it is a slow and tedious business to check the oil level in the sump.

Like all air-cooled engines, the Isetta's power unit heats up rapidly after a cold start, and quickly dissipates its heat after it has been stopped. Therefore, the choke control (placed conveniently close to the hand-brake lever) is frequently used, for short periods only, when starting. The combined dynamo-starter works well and we found starting simple and sure.

When the vehicle is parked for appreciable periods it seems advisable to use the small control immediately behind the top of the seat-squab; this is a three-position tap (on, off, reserve) for the fuel supply to the engine, The carburetter can tend to flood on occasion if the supply is not cut off when the vehicle is parked.

Although it deals with left-hand drive versions and is inevitably complicated through being tri-lingual (English, French, Spanish), the instruction manual is well-illustrated and provides the owner with all the information that he is likely to require. We think this is a particularly good point, for, as will be mentioned

The engine is positioned immediately ahead of the off-side rear wheel.

again later, owners of small vehicles like the Isetta frequently develop an affectionate regard for them and like to be well acquainted with their mechanisms.

The power of the headlights is amply sufficient to

The crab-track is evident in this view — also the wide rear window, and, immediately underneath it, the petrol filler.

enable the Isetta's full performance to be used at night on suitable roads. Strangely, the wiring of the switch operated by the steering-column lever that controls the main/dip beams is such that the main beams are operative when the lever is *down* and the dipped beams when it is *up*. Down-for-dipped is the more usual and logical arrangement. Any owner who did much nocturnal motoring and who felt strongly about the matter could, no doubt, quite easily change the wiring of the switch. But most owners will probably dismiss this as a minor idiosyncrasy on a par with the fact that the speedometer needle moves in an anti-clockwise direction as it records increasing speed.

Four forward speeds, good road-manners, ease of control and extremely modest dimensions (7 ft. 6 ins. long, 4 ft. 6 ins. wide) make the Isetta as rapid a means of transport over crowded roads as anything on four wheels. It can worm its way through city traffic, making use of gaps and opportunities that offer no possibilities for larger and less nippy vehicles.

The accelerator pedal calls for a definite second-pressure to be overcome when it is depressed beyond the first two-thirds of its travel. Restricting maximum speed to 45 m.p.h., and generally using only the first-pressure range of the accelerator — but nevertheless making full use of the indirect ratios — we obtained nearly 60 m.p.g. during motoring that included considerably more in-town than open-road driving.

Other tests which we carried out revealed that if it is driven with absolute maximum fuel economy in mind the Isetta can achieve better than 80 m.p.g.

Maximum speed was found to be 56 m.p.h. The manufacturer recommends 53 m.p.h. as the maximum cruising speed, and the sturdy, simple 298 c.c. engine can no doubt sustain this without undue stress, but we found 40–45 m.p.h. to be a more agreeable speed to maintain on the open road. In the Isetta that was submitted for test, a faster rate of progress than 45 m.p.h. brought increases in noise and vibration that the driver soon found tiring.

Isettas have proved their ability to tackle long journeys and to travel "off-the-tar". We view these demonstrations as providing convincing proof of the sturdiness of the design, but not as representations of normal usage. Long journeys *can* be undertaken with very small vehicles, but they are likely to prove very tiring and tiresome. We did not attempt any long-distance motoring in the Isetta, nor did we drive it over corrugated roads.

It should perhaps be explained that we do not accept a vehicle for test unless — from our knowledge of its design, and, very often, from our experience of driving a similar model belonging to a friend or acquaintance — we judge that there is a strong probability of our being able to recommend it to a section of our readers. We believed that there was every likelihood of our being able to recommend the BMW Isetta 300 Motocoupe to motorists who seek (probably as a second-car) an inexpensive, attractive, reliable and very economical miniature vehicle for in-and-about-town and short-distance trips up to a maximum of about 30 miles "at a sitting".

A universal joint at the foot of the steering column enables the steering wheel to swing forwards when the door is opened. The door is spring-counterbalanced and opens easily.

Standard equipment includes an external rear-view mirror as well as an internal one.

Our test fully satisfied us of the accuracy of this appreciation, and we have no hesitation in saying that we know of no vehicle that is better suited to such motoring requirements.

Some owners, in their enthusiastic appreciation of their Isettas, may elect to undertake quite long journeys, and to tackle surfaces ill-suited to ultra-small wheels and a crab-track. The Isetta has proved that it can give quite a creditable account of itself in these conditions, but there are many other vehicles that are far better equipped to deal with them.

The paintwork and general finish are good in view of the Isetta's price of £328 (at coast).

There is a certain "friendliness" about a small car, and a vehicle the size of the Isetta has a positively chummy character that appears to be appreciated in particular by women drivers. During the time that we drove an Isetta we spoke to many owners of these vehicles and we discovered that they often viewed the Isettas with considerable affection. One young mother said of her Motocoupe: "Sputnik, as we now call her, is definitely one of the family — valued, respected and well-loved." ●

SPECIFICATION AND PERFORMANCE

BRIEF SPECIFICATION

Make ... ... BMW
Model ... Isetta 300 Motocoupe
Style of Engine Single-cylinder four-stroke. Air-cooled (fan).
Bore ... ... 2·83 ins. (72 mm.)
Stroke... ... 2·87 ins. (73 mm.)
Cubic Capacity 18·61 cu. ins. (298 c.c.).
Maximum Horse-Power 13 b.h.p. at 5,200 r.p.m. (Compression ratio 7·0 to 1).
Brakes... ... ... Hydraulic
Front Suspension Independent. Swinging arms; coil springs.
Rear Suspension Narrow, rigid axle. Quarter-elliptic leaf springs.
Transmission System Clutch and manually engaged gears. Final drive by enclosed chain.
Gear Ratios (overall) 1st 23·21 to 1
2nd 12·14 to 1
3rd 8·17 to 1
Top 6·1 to 1
Rev. 30·0 to 1
Overall Length ... 7 ft. 6 ins.
Overall Width ... 4 ft. 6 ins.
Overall Height ... 4 ft. 5 ins.
Turning Circle Approx. 26 ft.
Dry Weight ... ... 730 lbs.
Price ... £328 at Coast Ports.
£343 in Johannesburg.

PERFORMANCE

Acceleration 0-30 m.p.h. 12·5 secs.
0-40 m.p.h. 24·1 secs.
In top gear from a steady 20 m.p.h. to 40 m.p.h. ... ... 30·4 secs.
Maximum Speed ... 56·0 m.p.h.
Reasonable Maximum Speed in 3rd Gear ... ... 38 m.p.h.
Fuel Consumption ... 59·2 m.p.g.
Test Conditions. Sea level. Dry road. No wind. 90 octane fuel.

BMW ISETTA 300

"Bubble car" is what the British label this miniature type of automobile

The Isetta 300 is certainly a contender for an economy award. Mileage ranges up to 70 m.p.g.

SPECIFICATIONS

Maker: Bayerische Motoren Werke AG, Munich, Germany
Cylinders: One (rear engine)
Bore: 72 mm. (2.83 in.)
Stroke: 73 mm. (2.87 in.)
Capacity: 298 cc.
Compression ratio: 7 to 1
Valves: Overhead
Cooling: Air
Maximum horsepower: 13
Gearshift: At driver's side
Speeds: Four
Final drive: Chain
Steering: BMW reduction
Suspension front: Independent, coil springs
rear: Semielliptic springs
Brakes: 7 in. diameter, total lining area 49.9 sq. in.
Tires: 4.80 x 10
Battery: 12 volt
Seats: Two/three
Weight: 770 lbs.
Wheelbase: 58 in.
Length overall: 93.7 in.
Width overall: 55.5 in.
Height overall: 52.7 in.
Instruments: Speedometer
Turning circle: 24 feet
Ground clearance: 6 in.
Price: $1,098 (sunroof, $1,048)

ORIGINALLY the Isetta was powered by a two-cylinder, two-stroke engine. It now employs a single-cylinder, four-stroke power unit. Other changes include a bit less bulge at the rear, and more parcel space. The rear suspension also is new.

The front of the car is also the door (the only door) and the steering column swings in and out with the door. The steering column also is conveniently used as a handle with which to pull the door shut after entering the car.

The single seat is bench type, and can carry two adults and a child. Air ducts from the finned cylinder provide heating and demisting.

The starter is combined with the generator (Siba Dynastart). The gearbox is of the motorcycle type, and is controlled by a very handy gearshift lever. With the extremely narrow track of the two rear wheels (20 inches), there is no differential.

Cutaway view of the car. The seat will take two adults and a child. Parcel space isn't bad.

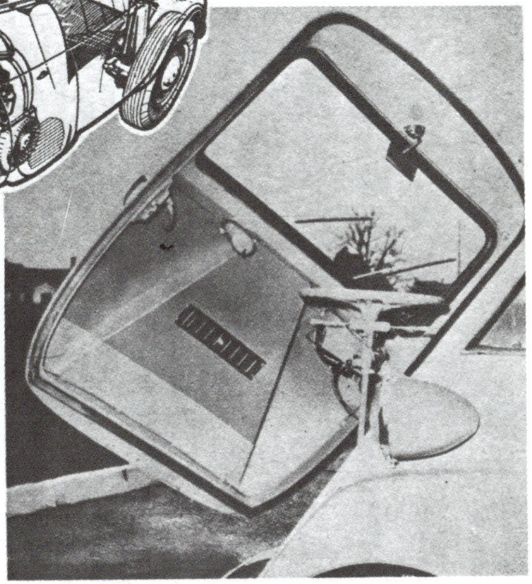

Right: And here is the famous front-opening door. The wheel is used as a convenient handle.

Roadtest Summary: PLUS—The big feature of the BMW Isetta 300 is its operating economy. If you are willing to stay at around 30 m.p.h. you can get a very comforting 70 miles per gallon or so. If you want to drive it hard—and you can—you may expect as high as 55 m.p.g. This is one of the dividends of the changeover a few years ago from the car's two-stroke engine to the present four-stroke.

The performance is lively if the driver utilizes the good gearbox—and it is easy as well as rewarding to use it to the full.

The car is sturdily built. Despite the nearness of riders to engine, if the car is driven at moderate speeds the noise is not irksome. The new springing in the rear makes for a comfortable ride, including really rough surface as well as parkway.

Roadholding generally is tops, and cornering gives practically no roll. The position of the engine at the right rear acts to balance nicely the weight of the driver at the front left.

The front-opening door makes entry and exit delightfully easy. Weatherproofing is equally fine. Visibility is excellent. Gear changing can be done as fast as you please. The engine starts at once—and almost without a sound. Needless to say, the car is a pleasure to park.

Acceleration: Zero to 30 m.p.h.—12 seconds; zero to 40 m.p.h.—23.5 seconds. Top speed, 50 m.p.h. Mileage, 55 to 70 m.p.g.

MINUS—While moderate speeds do not result in too much engine noise, it is a different story at around the car's top speed. High headwinds also do the performance and noise level no great good.

The steering, while light and precise, is disappointing so far as the turning circle is concerned, and 2.5 turns lock to lock would appear unnecessarily high for so light a car. •

BMW ISETTA 600

An interesting variation of two-door coachwork is this Isetta: one door in front, one on side.

SPECIFICATIONS

Maker: Bayerische Motoren Werke AG, Munich, Germany
Cylinders: Two, horizontally opposed (rear engine)
Bore: 74 mm. (2.9 in.)
Stroke: 68 mm. (2.68 in.)
Capacity: 582 cc.
Compression ratio: 6.8 to 1
Valves: Overhead
Cooling: Air
Maximum horsepower: 26 at 4,500 r.p.m.
Maximum torque: 28.9 lb./ft. at 2,000 r.p.m.
Gearshift: On floor
Speeds: Four (synchronized)
Rear-axle ratio: 5.43
Steering: Worm and nut
Suspension front: Independent, coil springs
rear: Independent, coil springs
Brakes: Total lining area, 64 sq. in.
Tires: 5.20 x 10
Battery: 12 volt
Seats: Four/five
Doors: Two
Weight: 1,166 lbs.
Wheelbase: 66.9 in.
Length overall: 115 in.
Width overall: 55.12 in.
Height overall: 54.12 in.
Instruments: Speedometer, ignition and oil-pressure warning lights
Turning circle: 26.2 feet
Ground clearance: 7 in.
Price: $1,398 (with sunroof, $1,487)

Detail and finish are of a high standard. That gearshift lever lies directly under one's hand.

Engine ventilation louvers serve as a tasteful decorative touch. The insignia is a famous one.

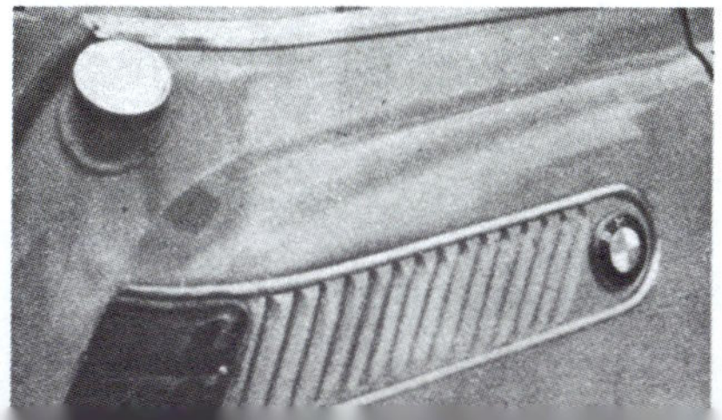

More than just a big version of the Isetta 300, this is a real motorcar

NEAT lines, amazing use of the available space, and good finish mark the BMW Isetta 600. Like the Fiat 600, with which it will inevitably be bracketed, it demonstrates that a "small" small car need not be ungainly, or handicapped by an appearance which proclaims "economy."

There are two doors; one opens the whole front end of the car, as on the Isetta 300, and there is another on the right side of the car. The doorless side makes for a graceful, unbroken panel which contributes much to the sleek look. The engine louvers are not only functional, but add a tasteful decorative note.

The floors are flat front and rear (in contrast to some other rear-engined cars which are needlessly marred by high "tunnels") and are covered with neat rubber mats. There are deep, useful pockets built into the side panels; together they hold more than even a good glove compartment would.

The headlining is of washable plastic; the upholstery is quite good-looking.

Front-window panels slide forward and back for ventilation and elbow room. The rear windows are fixed.

The front seats are well contoured, giving good support and making for comfort on extended journeys.

Roadtest Summary: PLUS—The first impressions one gets in driving the BMW Isetta 600 are: The car is surprisingly easy to get into; the engine starts at the first pressure on the ignition key; the floor-mounted gearshift lever is right under the hand where it should be, and its action is direct and positive.

On the road one soon finds that the steering is as light and sensitive as one could possibly wish. The seats are roomy and comfortable—their springing is neither too harsh nor too soft. The contouring in the front seats makes for added comfort. Two riders are accommodated in the rear without crowding each other.

The ride on good roads is up to the best standard for a car its size; bumpy roads can be felt, but do not toss the occupants around.

Visibility is good in all directions, especially forward, where it is unexcelled.

Leg room is good in front, tolerable astern.

Top speed, 70 m.p.h. Mileage, 45 to 50 m.p.g.

MINUS—The engine is far from quiet, though the sound is not annoying when one has become accustomed to its note. The steering is a shade too sensitive until one has become used to it. (A heavy touch in this respect leads to an alarming wander.) More room should be provided between the gas pedal and the sloping front door—a big-footed driver finds his toe scraping the latter. •

Right: Showing how steering column is jointed to permit the front panel to be opened outward.

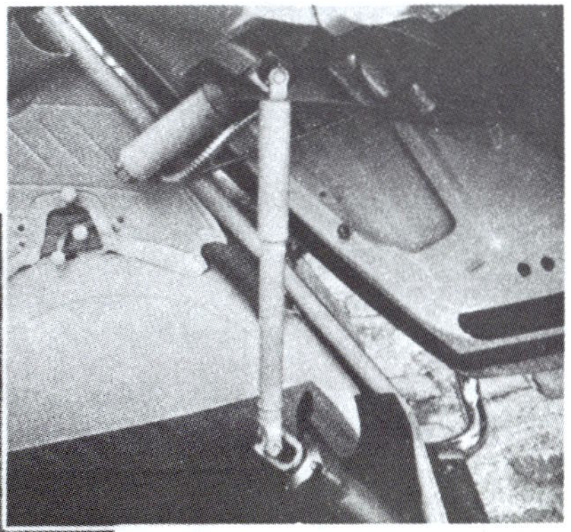

Below: With both doors open. Extremely clever use of room has provided good rear compartment.

Two-pedal Control for the Isetta

SMITHS SELECTROSHIFT AVAILABLE ON THE 300 PLUS MODEL

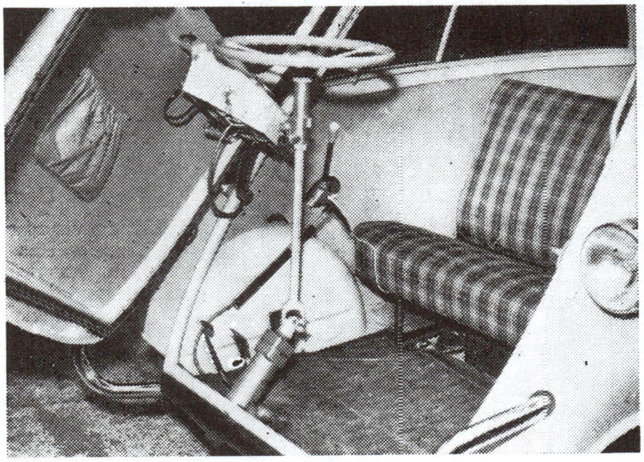

The Isetta's gear lever incorporates an electrical switch which breaks the clutch circuit for gear changing

FOR the first time a two-pedal control has been made available on a bubble car. The Smiths Selectroshift system, which has been under development for some considerable time, can be obtained as an optional extra on the Isetta 300 Plus model. It costs £16 13s basic, £24 19s 6d with purchase tax, bringing the total price of this model with two-pedal control to £282 8s basic and £424 19s including purchase tax.

The clutch, which is of the electromagnetic type, is fitted in place of the normal single-plate unit. There is no clutch pedal, and to move off from rest the driver merely presses the accelerator, and the clutch takes up the drive. For gear changing, the clutch is disengaged by the movement of a switch built into the gear lever and operated by the driver.

In construction, the clutch consists of a driving member attached to the rear of the crankshaft, a stationary member containing the windings of the coil, and an output member splined to the gear box mainshaft. The small gap between the driving and output members contains steel powder which is magnetized by the field built up by the coil, and the powder is attracted between the pole faces so that driving torque is transmitted. This arrangement is an improved version of the earlier Smiths clutches in which the coil rotates with the driving member. The latest arrangement gives a considerable reduction in the inertia of the moving parts, and the need for slip rings is avoided.

With the car at rest and first gear engaged for starting, the windings receive current from the dynamo; as the throttle is opened, the dynamo output rises and the corresponding increase in magnetic flux causes the clutch to take up the drive. Dynamo supply is modified by a resistance to give a suitable voltage-speed characteristic.

Clutch slip ceases at 12 m.p.h., when supply from the dynamo is cut off and current is taken from the battery instead. This is carried out by a governor driven from the speedometer gear box by a flexible cable and carried underneath the car. At the same time the dynamo output is returned to its normal characteristics for battery charging.

When gear changes are made, the switch in the gear lever is opened and the current to the windings interrupted, thereby disengaging the clutch. As the car is brought to rest, the governor contacts are remade at 10 m.p.h. and, as dynamo voltage falls, slip occurs and the clutch is disengaged. A small control box containing two relays and two resistors for the clutch circuits is under the seat.

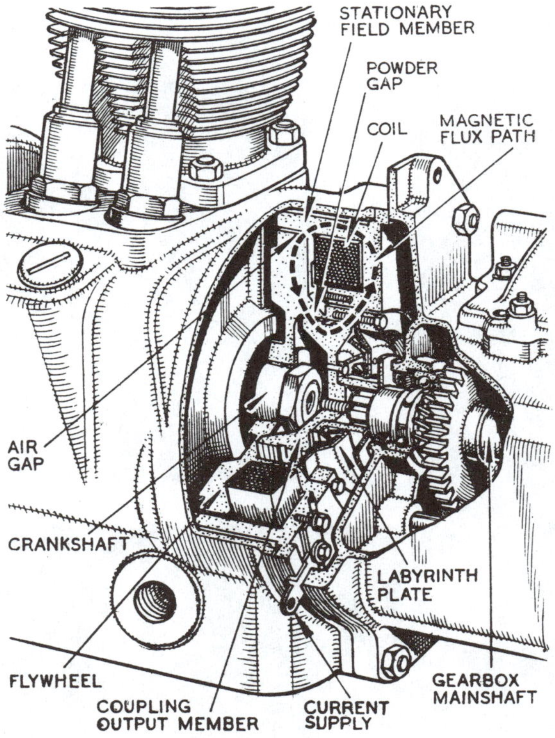

In the Smiths Selectroshift clutch, a magnetic field from stationary windings magnetizes a steel powder, by which the drive is transmitted

On the Road

Experience with a right-hand drive Isetta 300 fitted with Selectroshift showed that a big improvement in driving ease has been obtained, particularly in heavy traffic. Clutch engagement from rest is consistently smooth, however clumsily the accelerator is used. It is possible, though not recommended, for a getaway to be made in any gear, the duration of slip being longer, the higher the gear selected. As the car nears a standstill the clutch is not quite so smooth, there being a slight unevenness as it releases. There is never any difficulty in selecting first gear, a very small current being passed to the clutch coil to obtain slight relative movement between the dogs.

Both upward and downward changes can be made as fast as the lever can be moved, and it is almost impossible to make a noisy change. The ease of control is marred to some extent by the gear change mechanism, which is rather stiff and not sufficiently positive. Reverse is difficult to engage on occasions.

Those who like to brake with the left foot on cars with two-pedal control will find it impossible to do so on the Isetta, because of the position of the steering column. However, the clutch is sufficiently sensitive for low-speed manœuvring by means of the accelerator only. Restarting on a steep hill was entirely satisfactory, and it was even possible to run backwards with first gear engaged, press the accelerator to stop the car, and then return up the hill.

The Isetta 300 is manufactured in this country by Isetta of Great Britain, Ltd., New England Street, Brighton

AROUND THE WORLD IN THIRTY DAYS

. . . to see more minicars

BMW With a sort of pincer movement on the car market from their smallest economy model, the Isetta, to their most expensive sports car, the 507, BMW may be beginning to fill in the gaps with their latest economy miniature, the 600. A visual hybrid, it looks like an Isetta at the front and a Fiat 600 at the rear.

Entry for the front seats is through the front opening door, which brings with it steering wheel and instrument panel; for the rear seats, through one door on the right side. Head- and leg-room are surprisingly ample, front and rear; seats are comfortable and visibility all around is excellent.

A 19.5-bhp opposed two-cylinder engine, mounted behind the rear axle, is actually a de-tuned version of an air-cooled motorcycle powerplant and is exceptionally quiet and reliable. Chassis, with a ladder-type frame, is a beautiful study of automobile engineering on a miniature scale.

Driving the little BMW on a winding road illustrates the enormous progress made in building handling characteristics into rear-engined cars. There is definite understeer, no tendency for the tail to break loose, and only modest roll at near-sports-car speeds. Lack of pitching highlights a good ride. The car will pull in fourth gear from 14 to 60 mph with four people on board.

BMW ISETTA "600"

U.S. DISTRIBUTORS
FADEX COMMERCIAL CORP.
487 PARK AVENUE
NEW YORK 22, N. Y.

WEST COAST:
519 WEST 15th STREET
LONG BEACH, CALIFORNIA

BMW ISETTA "600"

EXCLUSIVE U.S. DISTRIBUTORS

FADEX COMMERCIAL CORP.
487 PARK AVENUE
NEW YORK 22, N. Y.

WEST COAST:
FADEX WESTERN MOTORS, INC.
519 WEST 15th STREET
LONG BEACH, CALIFORNIA

EXCLUSIVE U.S. DISTRIBUTORS

FADEX COMMERCIAL CORP.
487 PARK AVENUE
NEW YORK 22, N.Y.

WEST COAST:
FADEX WESTERN MOTORS, INC
519 WEST 15th STREET
LONG BEACH, CALIFORNIA

BMW ISETTA "600"

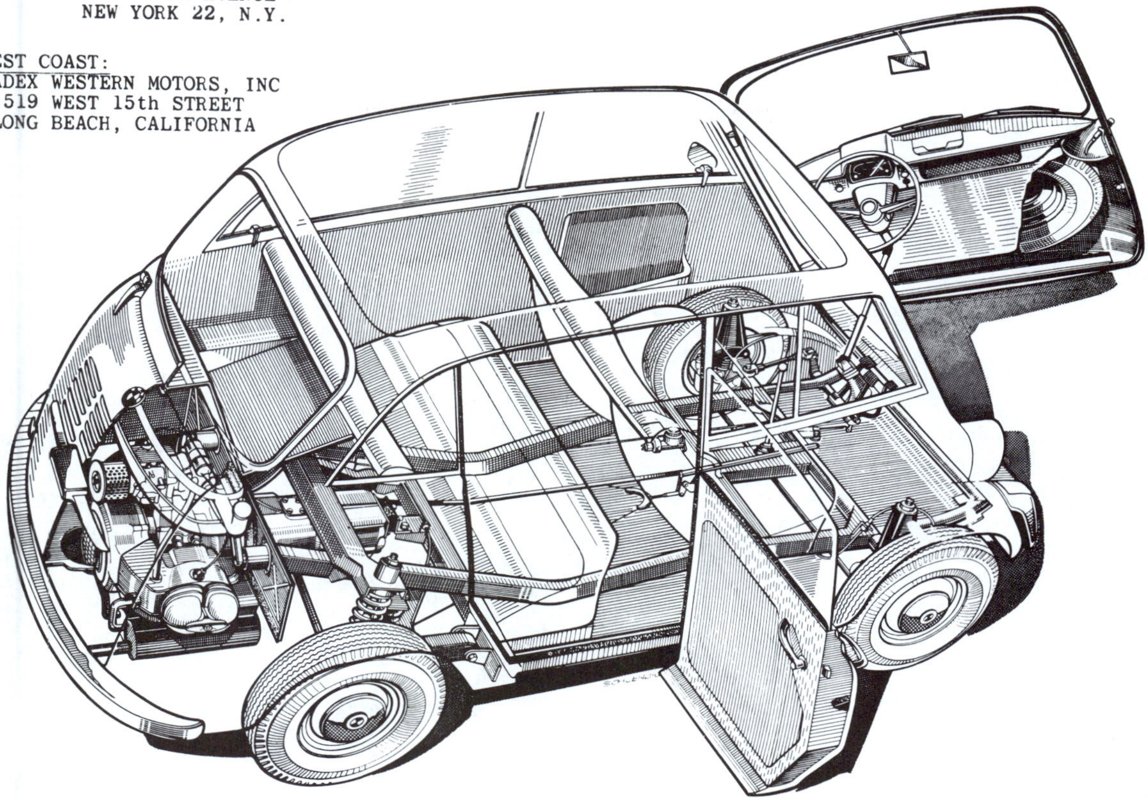

BMW ISETTA "600"

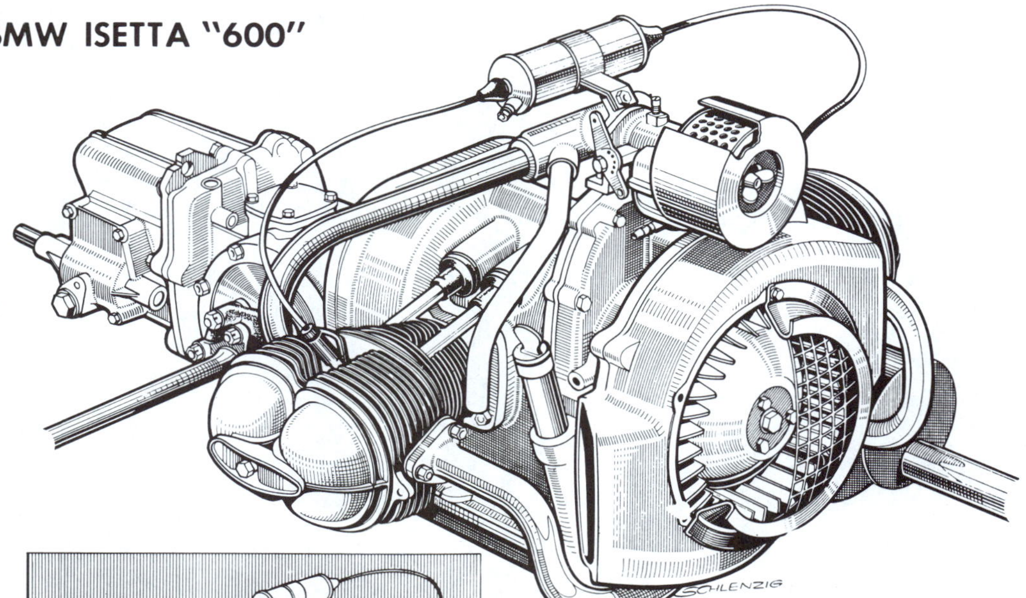

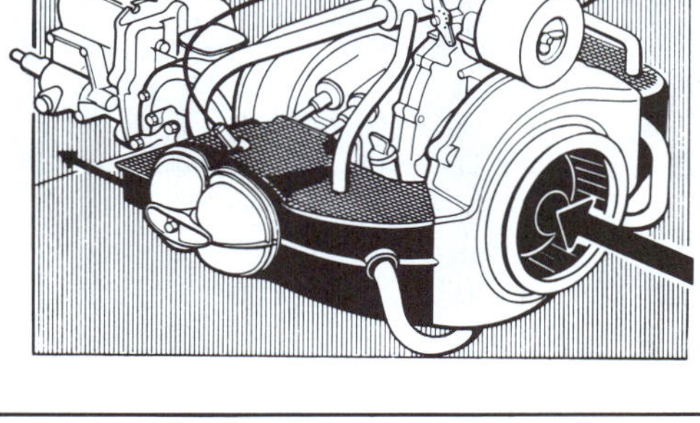

2 CYLINDER AIR-COOLED ENGINE

EXCLUSIVE U.S. DISTRIBUTORS

FADEX COMMERCIAL CORP.
487 PARK AVENUE
NEW YORK 22, N. Y.

WEST COAST:
FADEX WESTERN MOTORS, INC.
519 WEST 15th STREET
LONG BEACH, CALIFORNIA

CURRENT MODEL : BMW ISETTA "300"

Who else...but Cary Grant emerging from his BMW Isetta 300
on Park Avenue in New York City in 1958.

Bewundert
am
BMW
Sechshundert
Motor und Sport
ADAC
Hannoversche Presse
Der Stern
AUTOMOBIL Revue
hobby
DER TAG
Volle Einigung
in der Saarfrage
Radmarkt

Der neue BMW 600 im Urteil der Presse

Selten wohl erhielt ein neues Fahrzeug von der Presse so viele vorbehaltlose Anerkennungen wie der neue BMW 600. Was dabei gerade für Sie sehr aufschlußreich sein dürfte, ist die Tatsache, daß sich die anerkennenden Urteile nicht nur auf rein konventionelle Vorteile — die bei einem BMW-Wagen ohnehin selbstverständlich sind — beschränken. Im Brennpunkt des Beifalles der Fach- und Tagespresse stehen vielmehr besonders jene neuartigen Konstruktionsprinzipien, die bisher noch nicht gekannte Fahreigenschaften und Raumverhältnisse ermöglichen, also jene markanten Merkmale, in denen sich der BMW 600 von anderen vergleichbaren Wagen unterscheidet. Bitte lesen Sie, was die bekanntesten Fachjournalisten schreiben. Wir sind überzeugt, daß auch Sie nach diesen objektiven Pressestimmen und einer Probefahrt — zu der Sie übrigens jeder BMW Vertragshändler gern und unverbindlich einlädt — von dem BMW 600 begeistert sein werden.

Das Auto, Motor und Sport

Er besitzt verschiedene eindrucksvolle Merkmale und Eigenschaften, die in dieser Zusammenstellung nur ihm zu eigen sind. Noch vor wenigen Jahren hätte man es kaum für möglich gehalten, daß an einen Kleinwagen Forderungen gestellt werden könnten, wie sie etwa dieser neue BMW erfüllt.

Kölnische Rundschau

Überhaupt sind beim BMW 600 die verschiedenen Innenmaße so geschickt ausgelegt, daß sie die vieler anderer Kleinwagen übertreffen und oft sogar die Maße eines Mittelklassewagens erreichen. Das Maximum an Raumausstattung läßt sich nicht durch Konstruktionsprinzipien erreichen, die vom mittleren oder großen Wagen übernommen sind. Hier müssen ganz neue Wege beschritten werden. BMW hat sie mit dem BMW 600 bereits eingeschlagen.

Auto

Der BMW 600 ist ein gelungener Wurf, weil er nicht nur fahrtechnisch und wirtschaftlich viele Vorteile bietet, sondern vor allem so zweckmäßig ausgestattet ist, wie es gerade die Interessenten sich wünschen.

Frankfurter Nachtausgabe

Der BMW 600 wurde auf den sich über 14 Tage erstreckenden Probefahrten in keiner Weise geschont. Meine allein schon auf Grund des verwendeten Motors hochgespannten Erwartungen wurden mehr als erfüllt. Der robuste Boxermotor von 20 PS, der in der bisherigen Seitenwagenmaschine eine Leistung von 35 PS hatte, sollte durch seine aus Versicherungsgründen vorgenommene Drosselung eine Lebensdauer von 100 000 Kilometern garantieren. Er kommt mit seinem vollsynchronisierten und leicht zu schaltenden Getriebe überraschend schnell auf eine Spitzengeschwindigkeit von 100 km/st, ja sogar noch darüber. Bereits im zweiten Gang, der selbst im Gebirge kaum benutzt zu werden braucht, läuft er über 60 Kilometer. Im dritten Gang schafft er mit einer Vehemenz, wie man sie selbst bei mittelstarken Wagen nur selten antrifft, 85 km/st.

ADAC Motorwelt

Das Triebwerk war seit Jahrzehnten vorhanden, und zwar in Gestalt des heute noch in der ganzen Welt als führend geltenden Zweizylinder-Boxermotors mit angeblocktem Getriebe. Der Schwerpunkt im konstruktiven Komplex eines Kraftfahrzeuges liegt wohl im Motor. Dieser neue Kleinwagen hat eines der besten und meistbewährten Triebwerke dieser Hubraumklasse überhaupt. Was aber besonders interessiert, ist die Karosserie. BMW hatte schon vor Jahrzehnten den Werbeslogan „Innen größer als außen". Wenn je dieses Schlagwort zutrifft, dann für den BMW 600. Ob wir den Motor, das Fahrwerk oder den Aufbau betrachten, überall entdecken wir die traditionelle BMW-Qualität. Daß man mit diesem famosen Fahrzeug zu einem Preis unter der 4.000-DM-Grenze kommen konnte, ist eine Leistung, die wohl hauptsächlich darauf zurückzuführen ist, daß die Entwicklung des Triebwerkes keine großen Sorgen mehr bereitete.

Stuttgarter Nachrichten

BMW schuf für das Geld, das man selbst für ein anspruchvolles „Mobil" anlegen mußte, einen Wagen, der dem zweifellos mehr und mehr zunehmenden Luxusbedürfnis auch des weniger mit materiellen Gütern gesegneten Käufers entgegenkommt. Das Aufregendste ist zweifellos die Karosserie. Man muß bloß den Klang der schließenden hinteren Türe hören und weiß — Qualität.

Austro Motor, Wien

Wie nicht anders zu erwarten, brachte das Werk auch diesmal etwas, das der hervorragenden Rolle seiner Motorräder und schnellen Autos ebenbürtig ist. Schlicht und einfach ausgedrückt, heißt das, BMW's Jüngster hat — soferne man dies durch eine kurze Probefahrt beurteilen kann — derzeit die beste Federung aller Kleinwagen und dazu eine Straßenlage, die mehr ist als nur gut. Dazu gibt es ein Getriebe, das selbstverständlich synchronisiert ist. Der Clou des Getriebes ist aber der erste Gang, den man ebenfalls synchronisiert hat. Der BMW Kleinwagen besitzt zum Unterschied von anderen Konstruktionen einen kräftigen, verwindungssteifen Stahlrohrrahmen.

Schwäbische Landeszeitung

Die Innenmaße des BMW 600 übertreffen teilweise die einzelner Fahrzeugtypen der

Mittelklasse, wodurch der Wagen vier Erwachsenen, selbst für lange Reisen, ausreichend Platz gewährt ... Selbst bei großer Geschwindigkeit ist der Motor kaum zu hören.

KFZ-Fachblatt
Die große Elastizität des Motors geht aus der Spanne zwischen der geringsten einwandfrei zu fahrenden Geschwindigkeit und der höchsten hervor; sie liegt beim 4. Gang zwischen 20 km/st und 100 km/st (!!). Den 1. Gang braucht man daher nur zum Anfahren, da aber alle vier Gänge sperrsynchronisiert sind, kann man „mit dem kleinen Finger" schalten, völlig mühelos herauf und herunter.

Berliner Verkehrsspiegel
Wenn man bedenkt, daß der gleiche Motor im Motorrad 28 PS und in der Sportausführung sogar 35 PS leistet, kann man sich ausrechnen, welche lange Lebensdauer und welchen ruhigen Lauf der Motor haben muß, wenn er, wie beim BMW 600 nur 19,5 PS abzugeben braucht. Selbst in den unteren Gängen und bei Belastung hört man vorn nur ein leises Summen.

das noch durchkommende Rollgeräusch ist verschwindend gering. Aber selbst auf der hinteren Sitzbank kann man ohne weiteres bestätigen, daß der BMW 600 der derzeit leiseste Kleinwagen ist! Der Einstieg ist auch zur hinteren Sitzbank außerordentlich bequem und schafft für Kleinwagen völlig neue Maßstäbe. Die Federung muß als hervorragend ausgelegt bezeichnet werden, ihre Dämpfung ist ebenfalls first class. Auch hier darf man ruhig sagen: der derzeit bestgefederte Kleinwagen seiner Klasse ... Der BMW 600 liegt mit einem Abstand von 1850 mm zwischen der Oberkante der hinteren Sitzlehne und den Fußpedalen, der einen objektiven Vergleichsmaßstab für die Sitzbequemlichkeit bietet, noch über dem Durchschnittswert eines Mittelwagens ...

Frankfurter Allgemeine Zeitung
Mit dem in allen Gängen vollsynchronisierten Vierganggetriebe konnten wir störungsfrei kreuz und quer schalten. Im vierten Gang werden 7 Prozent Steigung bewältigt, so daß praktisch auf der Autobahn schaltungsfrei gefahren werden kann.

8-Uhr-Blatt
Wenn die Bayerischen Motoren Werke in München jetzt ihren neuen BMW 600 vorstellten und wir diesen neuen Kleinwagen als „Fahrzeug ohne Kinderkrankheiten" bezeichnen, dann haben wir dafür unseren guten Grund. Der neue BMW 600 ist in all seinen wichtigsten Einzelteilen schon so vieltausendfach erprobt, daß von neuen auftretenden Schwierigkeiten keine Rede mehr sein kann. Die Maschine, als wohl das wichtigste Teilstück der Neukonstruktion, ist eines der Hunderttausenden von Exemplaren, die bereits auf allen Straßen der Welt gefahren wurden, und es war vor allem die Maschine, die den Weltruf von BMW mit gründen half.

Nürnberger Zeitung
Der Motor ist das jüngste Kind einer Ahnenreihe, die bis auf das Jahr 1923 zurückreicht und den länger als ein Jahrzehnt gehaltenen Weltrekord mit 280 km/st enthält. Der vordere Einstieg wird durch eine hintere Seitentür ergänzt, und dadurch erhält dieser Wagen — welche ideale Idee — als Zweitürer die Wirkung eines Viertürers.

... und so billig fahren Sie mit dem BMW 600

Wieviel kostet die Unterhaltung?

a) Gesetzliche Haftpflicht Versicherungssumme DM 100 000,—	jährlich	DM 120,—
	+ 5 % Vers. Steuer	DM 6,—
	Summe	DM 126,—
b) Kraftfahrzeug-Steuer	jährlich	DM 87,—
Beständige Kosten aus a) und b)	jährlich	DM 213,—
	monatlich	DM 17,75
	wöchentlich	DM 4,43
	täglich	DM 0,63

So billig sind die Fahrkosten je 100 km:

I. Kraftstoffverbrauch 4,5 Ltr. à DM 0,66		DM 3,02
II. Ölwechsel 2 Ltr. auf 2000 km à DM 2,80		DM 0,28
III. Bereifung		DM 0,45
IV. Pflegedienste und Wartung		DM 0,60
	je 100 km	DM 4,35
	je 1 km	DM 0,04
Bei Besetzung mit vier erwachsenen Personen	je Person	DM 0,01

Ebenso günstig ist Ihre Steuerersparnis, wenn Sie mit dem BMW 600 von Ihrer Wohnung zur Arbeitsstätte fahren. Dabei ergibt sich folgende Vergünstigung:

	Brutto-Einkommen je Monat			
	DM 450,—	DM 520,—	DM 650,—	DM 780,—
Entfernung zur Arbeitsstätte:	monatliche Steuereinsparungen:			
10 km	DM 15,75	DM 18,50	DM 22,30	DM 24,95
20 km	DM 21,85	DM 34,85	DM 45,80	DM 52,90
30 km	DM 21,85	DM 34,85	DM 62,15	DM 76,40
40 km	DM 21,85	DM 34,85	DM 62,15	DM 92,75

Automobil-Revue
Als kleine Sensation wirkt die bei einem Fahrzeug dieser Klasse erstmalige Anwendung eines vollsynchronisierten Vierganggetriebes.

Der Tag, Berlin
Sehr überzeugend ist die Radaufhängung und Federung gelöst worden. Bei einer Probefahrt konnten wir uns von den erstaunlichen Fahreigenschaften dieses kleinen Fahrzeuges überzeugen. Man freut sich auch über die anatomisch richtig geformten Sitze, über den relativ großen Innenraum und — über die für einen Kleinwagen ungewöhnlich leise laufende Maschine. Neue Feststellung: Welchen Kleinwagen konnte man bisher im direkten Gang ruckfrei von 20 km/st hinaufbeschleunigen. Für den Techniker dürfte dieses Fahrzeug eine „technische Delikatesse" sein.

Radmarkt, Bielefeld
Sitzt man hinter dem Lenkrad, hört man den Motor überhaupt nicht mehr, und auch

Neckar-Echo
Mit der geräumigen und gefälligen Karosserie, sowie der sorgfältig abgestimmten Motorcharakteristik wurde ein neues Kleinfahrzeug geschaffen, das in seiner Fahrleistung und im Komfort an der Spitze aller zur Zeit auf dem Markt befindlichen viersitzigen Kleinwagen liegen dürfte.

Abendzeitung, München
Neutrale Fachleute vertreten die Ansicht, daß der BMW 600 wohl das solideste ist, was zur Zeit auf dem Kleinwagensektor geboten wird. Der BMW 600 ist geräumig, wirklich geräuscharm (!), sehr gut gefedert, hat eine gute Straßenlage und beschleunigt ausgezeichnet.

Allgemeine Zeitung, Mainz
Komfort und Fahrleistung stellen den BMW 600 in den Vordergrund der jetzt neu auf den Markt kommenden Kleinwagen ... Bei Raumverhältnissen eines Wagens der Mittelklasse stattete BMW seinen neuen Kleinwagen vom Typ 600 mit einer Fülle technischer Errungenschaften der Gegenwart aus.

Der Stern
... Das ist bei dieser Größenklasse eines Automobils der vernünftigste Einstieg überhaupt.

... Unter allen Kleinwagen scheint er mir der am liebevollsten ausgestattete zu sein ... Das Motorengeräusch ist angenehm. Und die Luftkühlung ist das erste Kühlungsgebläse, das man praktisch nicht hört.

Hobby — das Magazin der Technik
Die Voraussetzungen für unsere Fahrt mit dem BMW 600 waren nicht gerade günstig. Gegen zwei widrige Umstände hatte er anzukämpfen: Einmal zeigte der Kilometerzähler erst 400, zum anderen herrschte ein bösartiger, böiger Wind. Aber beide Handicaps meisterte das Fahrzeug mühelos. 100 km/h — als Mittel mehrerer Messungen — schaffte der Wagen, ohne daß sich die schräg anfallenden Windböen besonders bemerkbar machten.

Technische Daten des BMW 600

Motor

Typ	BMW Boxermotor, 4 Takt, hängende Ventile, über Kipphebel und Stoßstange betätigt
Zylinderzahl	2
Hubraum	582 ccm
höchstes Drehmoment	3,9 mkg 3000 U/min
Motorleistung	19,5 PS 4500 U/min
Bohrung	74 mm
Hub	68 mm
mittl. Kolbengeschwindigkeit	10,2 m/S bei Höchstleistung
Verdichtungsverhältnis	1 : 6,8
Lage im Fahrzeug	Heck
Schmiersystem	Ölumlaufschmierung mit Hauptstromfilter
Kühlung	Luft mittels Gebläse, Gebläserad auf Kurbelwelle
Zündung	Batterie-Zündung
Anlaß-Lichtmaschine	12 V/130 W
Kraftstoffbehälter	20 Liter und 3 Liter Reserve, Lage im Heck, Betankung von außen
Kraftstofförderung	Gefälle
Vergaser	Drosselklappenvergaser mit Beschleunigungspumpe

Kraftübertragung

Getriebe	mit Motor verblockt, 4 Vorwärts- u. 1 Rückwärtsgang
Übersetzungsstufen	1. Gang 3,45; 2. Gang 1,94; 3. Gang 1,27; 4. Gang 0,846; Rückwärts-Gang 3,45
Synchronisierung	alle Vorwärtsgänge sperrsynchronisiert und schrägverzahnt
Schalthebel	in Fahrzeugmitte vor den Sitzen
Kupplung	Einscheiben-Trockenkupplung
Radantrieb	Kegelräder und Differential; Übersetzung: 1 : 5,43 Hinterachswellen mit doppelten Gummigelenken

Fahrwerk

Rahmen	Stahlrohrchassis aus verschweißten Längs- und Querträgern
Radaufhängung	vorn: spur- und sturzkonstante geschobene Längsschwingen in Parallelogrammanordnung hinten: breitgelagerte gezogene Längsschwingen, Lagerung in Gummi
Federung	vorn: Schraubenfedern auf Achsträger abgestützt; Federweg: 120 mm hinten: Schraubenfedern auf Rahmen abgestützt, progressive Wirkung durch voluminöse Gummipuffer, Federweg: 160 mm 4 Teleskopstoßdämpfer
Lenkung	Spindel-Lenkung; ungeteilte Spurstange; mitgelenkte Vorderfederung; von Federung unbeeinflußte, kinematisch richtige Lenkgeometrie
Reifen	5,20—10; Reserverad: hinter Vordertürverkleidung
Spur	vorn: 1220 mm, hinten 1160 mm
Radstand	1700 mm
Bremsen	Fußbremse hydraulisch auf alle 4 Räder, Handbremse mechanisch auf Hinterräder

Karosserie

Art	Stahlblech, auf Rahmen aufgesetzt
Türen	Fronteinstieg zur vorderen Sitzbank; beim Öffnen der Tür mitschwenkende Lenksäule; Seitentür rechts zum Einstieg zur hinteren Sitzbank
Sitze	mit Auflagepolstern anatomisch richtig geformt; Form der Rücklehne gerundet wie bei Sesseln; hintere Sitzlehne umlegbar; hintere Sitzbank leicht demontierbar zur Schaffung eines Laderaumes; beide Sitzbänke leicht herausnehmbar
Gepäckraum	hinter Rücklehne der hinteren Sitze; Inhalt: 100 Liter (gemessen bis Heckscheiben-Unterkante)
Heizung	kombinierte Frischluft-Auspuffheizung

Elektrische Anlage

Batterie	12 V, 24 Ah
Frontleuchten	hochliegende Scheinwerfer 130 mm ⌀ 45/40 W in Karosserie einbezogen, neuartiges asymmetrisches Abblendlicht
Heckleuchte	kombinierte Schluß-Brems-Blinkleuchte
Fahrtrichtungsanzeiger	Frontblinker in Stoßstange, Heckblinker in Heckleuchte; automatische Rückstellung
Bedienungsschalter	für Blinker u. Abblendung unterhalb des Lenkrades
Lichthupe	unterhalb des Lenkrades

Abmessungen

Länge	2900 mm
Breite	1400 mm
Höhe	1375 mm
Gewicht	fahrfertig 565 kg; zulässiges Gesamtgewicht 900 kg
Wendekreisdurchmesser	ca. 8 m
Höchstgeschwindigkeit	100 km/st
Steigfähigkeit im 4. Gang	ca. 7%
Kraftstoffnormverbrauch nach DIN 70030	5,5 Ltr./100 km
Ölverbrauch	0,1 Ltr./100 km

Änderung der Konstruktion und Ausstattung vorbehalten.

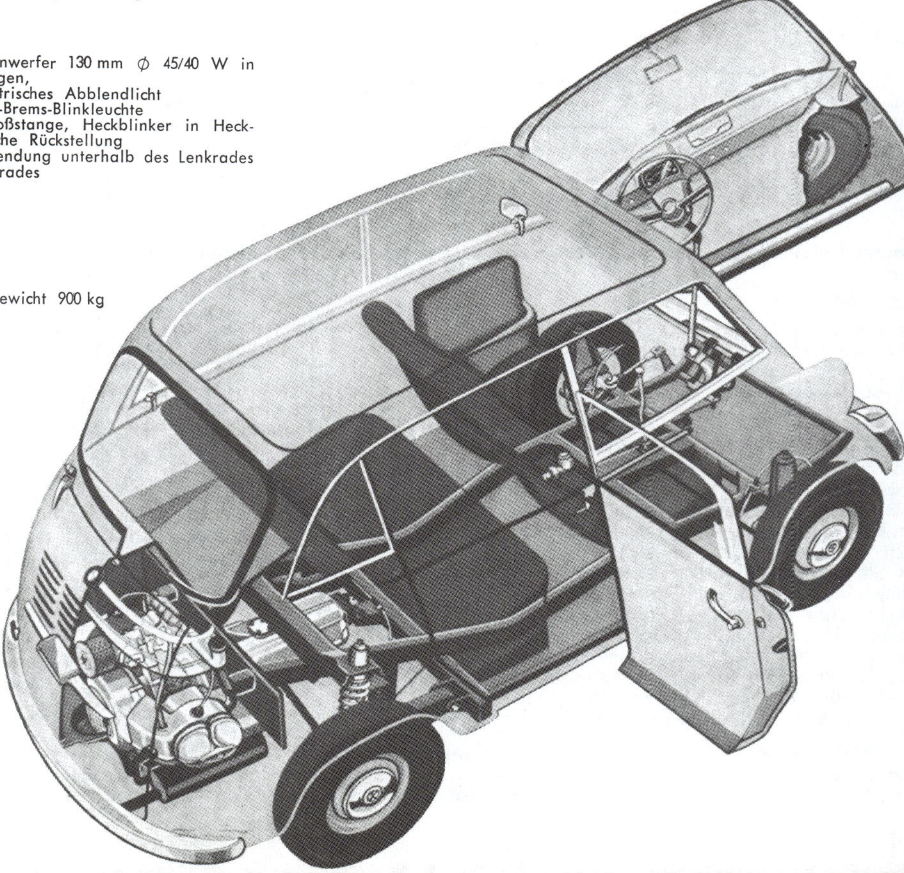

BMW 503 Convertible Coupe

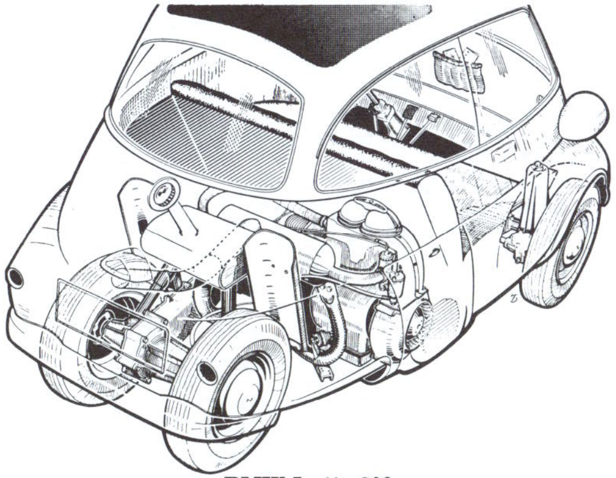

BMW Isetta 300

BMW (German)

Best known in this country for its Isetta miniature cars (a favorite premium on television programs), the BMW (for Bayerische Motoren Werke) also builds large, comfortable automobiles with V8 engines.

The BMW 503 convertible coupe even has power-operated top and windows. The V8 has 193 cubic inches, half the displacement of the new Edsel V8. Yet it has a top speed of about 118 miles per hour and an average fuel consumption of about 21 miles per gallon.

It seats four and has adequate luggage space in the trunk.

With slightly more power, but the same basic V8 engine, the 507 Touring Sports Car has a top speed of about 140 miles per hour, average consumption of 24 miles per gallon.

New this year (and not yet in production) is the BMW Isetta 600 which will seat four. It has two doors, one in front, one on the side for the rear seat. Four-wheel independent suspension, a horizontal-opposed two-cylinder engine of about 35 cubic inches make the car a versatile, practical machine approximately 9½ feet long. Fuel consumption should be near 50 miles per gallon.

BMW 507 Touring Sports Car

The new version of the Isetta, introduced this year, is the BMW 600. Drawing at right shows the layout of the four-passenger car. It has the same front-opening door as the Isetta 300, but in addition has a side-opening door for the rear seat passengers.

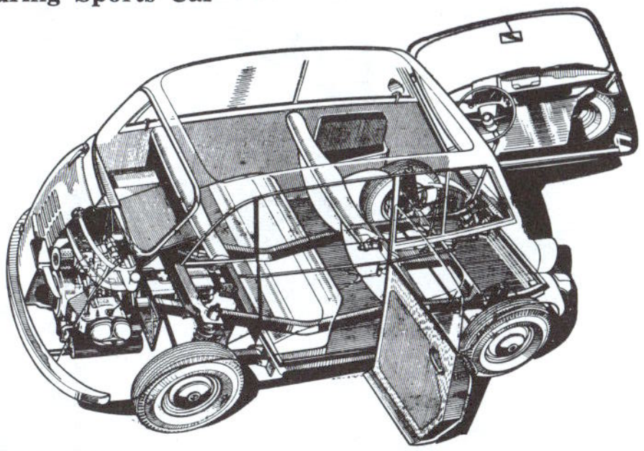

BMW Isetta 600

BMW ISETTA 300

from the makers of one of the world's finest Cars comes this amazing "second" car

by DICK KNELL

With the Near East crisis, Europe was flooded with a sudden influx of economy cars, some conventional in design, some radical, some practical and some impractical. All were attempting to solve the problem of transporting two or more people with a maximum fuel economy. One of the more radical, yet extremely practical answers to the problem was the amazing BMW Isetta "300."

The Isetta has a very rigid box section frame with a tubular extension at the front that follows the body profile and could easily act as a roll bar in the event of a flip. The frame narrows drastically at the rear because of the very narrow rear tread which is necessitated by the lack of a differential. If the rear wheels were set farther apart there would be a considerable amount of wheel scrub when going around a corner, which would lead to extreme tire wear.

The workmanship and quality of materials in the Isetta is amazing, especially considering the price of just over a thousand dollars. The quality of the car puts to shame the best the United States has to offer, regardless of price, and is equal to any except the highest priced imported cars. There are no rough spots in the car and every piece fits perfectly.

The most novel feature of the car is the all-in-one door in the front of the car. The use of one door cuts the cost of this item in half as compared to conventional methods and, at the same time, offers the easiest method of entering and exiting we have yet seen. All that is necessary is to open the door, which then opens to its fullest, by itself, taking the instrument panel and steering column with it. This offers an uncluttered entrance, by folding the seat forward and pulling it out with one hand—the small tire and 10 inch wheel make it very easy indeed. The package shelf, behind the seat, is large enough to carry one suitcase and assorted small items. A luggage rack is offered as an optional extra for those who are contemplating a short trip.

Starting is immediate in all weather conditions and the choke is needed only to start. The engine warms up almost immediately and there is no need to warm the car up for more than a few seconds. The standard equipment folding top, which is one of the tightest fitting we have seen, is a boon indeed, for the quarter windows do not give adequate ventilation in warm weather and it is necessary to have the top open at all times.

The engine has the usual unfavorable high noise level usually found in air-cooled engines, however it does not become objectionable and, in fact, sounds rather pleasant at high speeds. The engine thrives on revolutions, consequently it becomes second nature to use the maximum speeds in the gears at all times. Although the engine will lug down quite well it is not recommended that you do so, primarily because the gas mileage suffers in the process. The BMW engine has a reputation for reliability and long life so it is safe to use all of the available revs.

The Isetta is not capable of keeping up with traffic for the first 100 yards but within a city block the car is up to 40 mph and is either even with traffic or ahead of it. Driving the Isetta in town is truly a revelation, for it feels like a big, full size car yet its dimiunitive size makes it possible to skim in and out of tight traffic jams with ease. It is possible to park the car in a space

allowing you to walk right in and sit down. The seating position is excellent, the annoying feature of the car is evident at once. The emergency brake handle is just to the left of the drivers seat and your leg constantly rubs against it, cracking into it every time you hit a bump.

The extreme forward placement of the seat and the location of the door make a rear engine location necessary. In order to keep the wheel-base to a minimum the engine is set on the side of the car. Access is by a small panel that unscrews from the body panel and offers only a minimum of accessibility. Removal of the engine, however, is a very simple matter.

The instrument panel consists of a speedometer, a light switch, starter and nothing more. The electric windshield wiper switch is on the door and there is a switch behind the seat on the parcel shelf for the gasoline reserve, which, when switched on, will carry the car about 40 miles to the next gas station. The gas tank holds 3½ gallons. The controls all function very admirably, with the exception of the heater which is lacking in heating ability. The headlights are extremely powerful and quite capable of handling at least twice the speed of the Isetta. The dimmer switch is mounted on the steering column and is much more convenient than the usual floor mounted button. The only unorthodox control is the gear change which is on the drivers left, keeping the floor clear of obstruction. The shifting action is very smooth and in only a short time even the automatic transmission bred women can shift with ease. The battery and tool kit are under the seat while the spare tire is placed behind the seat and can be reached

two inches longer than its 93.7 in. If even a space this small is not to be found then merely drive head into the curb, open the door and step right out onto the sidewalk.

It is often necessary to take avoiding action whenever you approach a hilly section with some small cars, not so with the Isetta. On any but the steepest hills it was necessary to slow down only for big American cars that were taking up the road at 25 mph. Essentially, the Isetta is a town car but it can be taken on trips if there is no great rush. The Isetta cruises happily at 50 mph and is not affected by side winds or disturbing road conditions as are some cars, but is as rock steady as a 4000 lb. car. Over roads with ruts or bumps the short wheelbase and the small tires cause a pitching motion to build up and caution must be used when approaching a corner gutter at high speed—the car won't leave the ground—you leave the car.

The Isetta is an ideal second car for the husband to drive to work or the wife to go shopping in. Its low initial cost, its low maintenance cost and its ease of operation make one stop and wonder about the tiny housewife who drives a huge American behemoth to go shopping for 10 lbs. of groceries. With the horrible traffic conditions of today and the almost total lack of parking space the Isetta seems to be the perfect answer.

SPECIFICATIONS

ENGINE	air-cooled single cylinder OHV
BORE	72 mm.
STROKE	73 mm.
CAPACITY	293 cc.
COMP. RAT.	7.0 to 1
H. P.	13 at 5200 rpm
TRANS.	four speeds ratios N.A.
FINAL DRIVE	enclosed chain
TIRE SIZE	4.80-10
BRAKES	7 in. dia.
TRACK, FRONT	47.2 in.
TRACK, REAR	20.4 in.
WHEELBASE	58 in.
OVERALL LENGTH	93.7 in.
CURB WT. (dry)	770 lb.
TOP SPEED	53.6 mph
ACCELERATION	adequate
GAS MILEAGE	64 mpg

Knowledgeable people buy Imperial
and they buy it by the case
WHISKEY BY HIRAM WALKER
IMPERIAL
LAND OF LINCOLN
303 014
19 ILLINOIS 58
IMPERIAL
HIRAM WALKER
Whiskey by Hiram Walker, the product of 100 years of distilling skill.
BLENDED WHISKEY • 86 PROOF • 30% STRAIGHT WHISKEY • 6 YEARS OR MORE OLD • 70% GRAIN NEUTRAL SPIRITS • HIRAM WALKER & SONS INC., PEORIA, ILLINOIS

ROAD TEST
ISETTA 300

Although the Isetta is unique in appearance, its height of 52.7 inches is similar to current U.S. cars (the '58 Plymouth is 56.6 inches), while 58-inch wheelbase and 93.7-inch overall length create a striking contrast (Plymouth wheelbase, 118 inches; overall length, 204 inches). The Isetta's tubular bumpers are correctly positioned and do the job required, but the aluminum hub caps are useless after one brush with the curb.

THOSE WHO HAD the unusual opportunity of driving one of the original Italian-made Isettas, as we did some two years ago, may well feel that this is not the answer to their personal transportation problem. Since then, both a growing interest in operating economy and a rising standard of living (if Ford's Mr. Breech will excuse us) have resulted in the birth and development of at least 10 significant minicars in Europe. The desire to reduce the car budget and the need for an additional car (corresponding to any car at all for many Europeans) are, in our experience, also important to a considerable number of American families. In importing the greatly improved Isetta 300, BMW evidently agree with us.

Based on sales in the field immediately above the Isetta's, in which the Volkswagen, Renault Dauphine and Metropolitan have made names for themselves, buyers seem to like something that is different as well as competent. Another year will tell us whether a conventional approach will pay off in General Motors' sales (Vauxhall and Opel) in a slightly higher price range; Volvo have already succeeded there.

In the ultra-economical group, so far represented in the U.S.A. by Goggomobil's sedan (the coupe is comparatively expensive to buy), the Messerchmitt and the test car in question, a designer is not so free to choose whether he will be conservative or radical. Considerations of ride, interior room and fuel mileage have to vie with reasonable construction costs before the package can get near the market. If overall appearance is a result rather than a determining factor—the reverse of Detroit practice—it only makes good sense.

That is evidently how the Isetta came to its present form. Personally, we think it is a grotesque attention-getter. Our test was conducted largely in rush-hour traffic, in the midst of a burst of "unusual" Southern California liquid sunshine, and we were glad that it did catch everyone's eye; we should have felt awfully vulnerable if it hadn't.

The front-opening door is a wise idea. Entrance around the double-jointed steering column is very easy for the driver if a passenger hasn't preceded him, and once he has learned not to grasp the wheel for support, for doing so will close the door smartly before he is ready for it. If it is raining (and it was) a considerable amount of water is likely to enter with him, but once in, he will be as snug as in any car of our acquaintance. The interior is so well sealed that the door cannot be slammed; a positive latch, quite a reach from the driver's position, must be engaged and can be locked.

Head-in parking is practical with a car whose length is approximately the width of an American car. In this position, one can step directly on or off the curb. Parallel parking presents a possible hazard, for once hemmed in, the driver can get neither in nor out. And as a final bit of nonsense before we leave the door, the car can be driven with it open in any

The gear lever shows at the left side of the wheel.

Forward entrance is adequate, provided the driver gets in first.

ROAD TEST ISETTA 300

position. We do not recommend it, nor do the manufacturers.

The plastic-covered seat is not deep, nor does it adjust to fit different-sized drivers. Adequate head room has been achieved at the cost of a corner post blind spot for a tall driver and somewhat restricted space for his legs and feet. These are no worse, however, than in many much larger cars. The passenger has plenty of room.

Instruments are basic only, consisting of a small speedometer with odometer, a generator discharge warning light, and two other lights indicating that the high beam is on and that the direction signals are operating. The headlights are dipped with a lever under the wheel, a pleasant system especially useful for signaling, and the light is blue, a German preference that we should like to see adopted here in place of red. As for the direction lights themselves, they are big and bright, and mounted high enough to be seen both front and rear.

The gearshift lever extends from the side panel at the driver's left. The pattern is an H, with synchromesh on 2nd, 3rd and 4th speeds. Fast shifts are smooth and enjoyable; a frequent reluctance to go into 1st speed is not at all enjoyable, particularly because a departure in 2nd is an ignominious thing indeed.

Also at the driver's left, the hand brake holds well. Small levers for the heater and the choke are exact duplicates and cunningly placed where they are easily confused in the dark. To be sure, very little choking is needed (particularly when one wants to turn on the heater instead). A big shelf extends all the way from the seat back to the rear window. Except for the gasoline reserve switch, it is available for many parcels

The enclosed final drive unit and narrow rear tread.

PHOTOGRAPHY: POOLE

The single-cylinder, air-cooled engine is at the right.

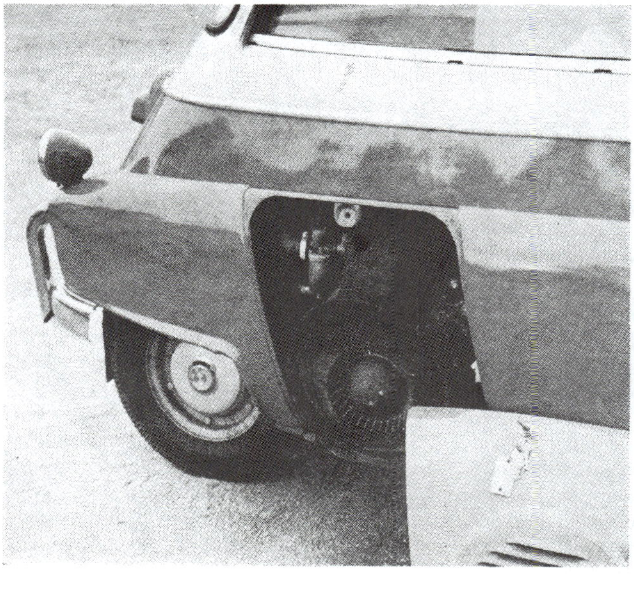

without blocking the 360° vision, if one doesn't mind their tendency to turn into missiles in a fast stop. Various pockets hold maps and such. Sliding windows on both sides, an optional air vent in the door, and a fabric roof insert (totally waterproof when closed) give excellent ventilation in hot or cold climates.

Quiet starting with a Siba Dynastart quickly gives way to a noise level that is less than pleasing. (Flintridge Motors, of Los Angeles, who supplied the test car, sell a polyurethane foam insulation kit that makes a surprising difference; it costs $30.00 installed.) There is no sense of strain from the single-cylinder, 4-stroke engine when it is revved up, but being right behind the seat, it is always evident. In spite of the unprepossessing acceleration figures, the 300 is like the smallest Berkeley (tested in last month's issue) in this one respect only: it gives a satisfying *sense* of acceleration, due partly to its size and partly to the willing response of the engine. Use of the upper 3 ratios of the gearbox is fairly constant, and no chore at all.

Here is yet another tiny car whose ride—at least, considering

the egg and we, tested on land and sea

size—is very good indeed. It is wise to watch for, and avoid, any large obstacles in the road. With the very narrow rear tread, this is not always easy, and the jounce is severe.

We had not expected so good an opportunity to try the Isetta in bad weather. On slick roads, it has a definite tendency to "wag its tail." This need be no problem if one knows what to expect, but novice drivers about to start their driving careers in an Isetta could benefit from a short driving course on a wet, empty parking lot.

Defensive driving, because of the Isetta's limited power and its dislike for really bad surfaces, is a necessity. We were once forced to take to a soft and unexpectedly deep road shoulder, and went through a bad moment or two. To give the little car its due, it showed no tendency to give up the ghost, but it did all the driving for a moment or two, while we merely looked on.

Gasoline mileage, of course, is almost incredible. The wisdom in changing from the original 2-cylinder, 2-stroke design to a single fairly large cylinder of 4-stroke pattern shows up especially in hard driving. That is the way it is fun to drive the 300, and it is pleasant not to have the miles-per-gallon figure drop as it can with a 2-stroke when it is pressed.

The apparent ruggedness of the design impressed us most favorably. There is very little to go wrong with the Isetta, and repairs should be both simple and cheap. The single spark plug can be reached from the driver's seat, and the carburetor is behind the slotted cover on the right side of the car.

The air-cooled engine is on the right to balance the weight of the driver, on the supposition that the car will frequently be driven without a passenger. We therefore tested acceleration with only the driver in the car, but found only slightly over a second's advantage compared to the ¼-mile figure in the data panel on this page. As usual, all figures in the data panel were computed with both driver and passenger.

For specialized uses—getting around a crowded, slow-moving city or a smaller place where traffic does not move at breakneck speed, or in isolated areas where service could be a problem for more complex machinery, and where economy is all important—there is no doubt that the Isetta is a vehicle to consider. It is well designed, well built, and it does its job efficiently.

ROAD & TRACK ROAD TEST NO. 159

BMW ISETTA 300

SPECIFICATIONS

List price	$1095
Wheelbase, in.	61.0
Tread, f/r	47.2/20.4
Tire size	4.80-10
Curb weight, lb	860
distribution, %	55/45
Test weight	1160
Engine	1 cyl, ohv
Bore & stroke	2.83 x 2.87
Displacement, cu in.	18.1
cu cm.	297
Compression ratio	7.00
Horsepower	13
peaking speed	5800
equivalent mph	61
Torque, lb-ft	14.0
peaking speed	4200
equivalent mph	44
Gear ratios, overall	
4th	5.54
3rd	8.18
2nd	11.9
1st	23.2

CALCULATED DATA

Lb/hp (test wt)	89.2
Cu ft/ton mile	51.5
Engine revs/mile	5700
Piston travel, ft/mile	2730
Mph @ 2500 ft/min.	54.9

PERFORMANCE, Mph

Top speed, avg.	49.7
best run	51.4
3rd (5750)	41
2nd (5750)	28
1st (6000)	15
see chart for shift points	
Mileage range	44/62 mpg

ACCELERATION, Sec.

0-10 mph	1.8
0-20 mph	4.1
0-30 mph	10.0
0-40 mph	20.0
0-50 mph (one way)	52.0
Standing start ¼ mile	27.7

TAPLEY DATA, Lb/ton

4th	110 @ 40 mph
3rd	180 @ 35 mph
2nd	270 @ 22 mph
1st	400 @ 12 mph
Total drag at 60 mph	N.A.

SPEEDOMETER ERROR

Indicated	Actual
10 mph	9.24
20 mph	18.6
30 mph	27.2
40 mph	36.0
50 mph	45.0
57 mph	51.4

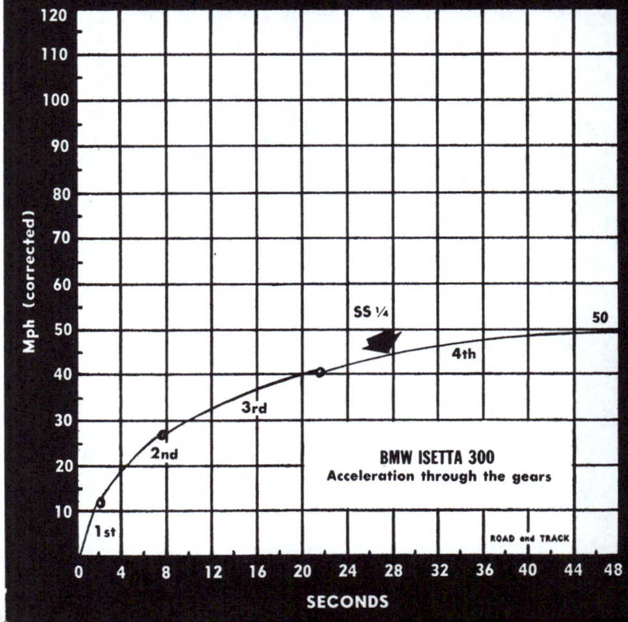

c'est une vraie voiture...

Sa banquette confortable et spacieuse, ainsi qu'un tableau de bord complet, lui assurent la même HABITABILITÉ et le même ÉQUIPEMENT, chauffage compris, que les voitures de très fortes cylindrées.

SA TENUE DE ROUTE est exceptionnelle, et ses PERFORMANCES étonnent plus encore aussi bien dans les moyennes enregistrées sur longs parcours, qui sont véritablement surprenantes, que dans la vitesse atteinte : 85 kms-heure.

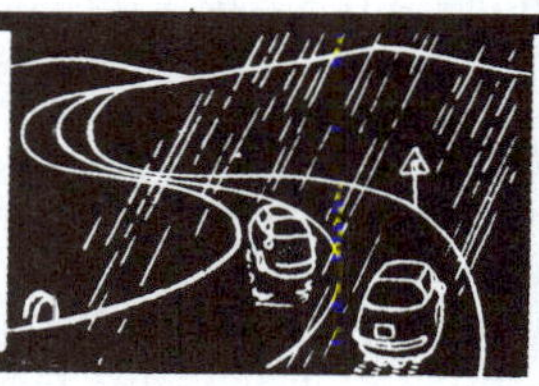

SON FREINAGE particulièrement efficace permet d'obtenir un maximum de sécurité dans le PILOTAGE en ville et sur route

Plusieurs valises, colis, ou BAGAGES peuvent aisément être placés dans le vaste espace qui leur est réservé à l'arrière de la voiture.

Dans un souci de perfection, un soin particulier a été accordé au GARNISSAGE INTÉRIEUR du véhicule qui souligne davantage encore l'ambiance de confort et d'élégance de la VELAM-ISETTA.

Pilotage encore facilité par l'utilisation d'une BOITE DE VITESSE dont la maniabilité émerveillera les utilisateurs.

elle offre en plus ...

UNE VISIBILITÉ PANORAMIQUE
inégalée jusqu'à ce jour,
encore accrue par le toit instantanément décapotable
qui en fait une voiture de tourisme idéale.

DES FACILITÉS DE STATIONNEMENT
qui font pâlir d'envie les automobilistes
constatant que deux ISETTA se logent aisément
à la place qu'ils occupaient avec leur seule voiture.

LA POSSIBILITÉ D'IGNORER PRATIQUEMENT LES EMBOUTEILLAGES
en raison de son stupéfiant
pouvoir d'infiltration.

UNE CONSOMMATION DE 3,5 L. AUX 100 KM
qui, alliée à son prix d'achat, fait de la VELAM-ISETTA
un moyen de transport non plus coûteux, mais économique.

UNE ÉGALE VALEUR
sur tous les parcours, en ville grâce à ses reprises,
et sur route grâce à ses performances exceptionnelles en côte.

UNE ACCESSIBILITÉ
par porte frontale,
qui séduira principalement les hommes de haute taille
et les femmes en robe du soir.

Construction entièrement française

Licence ISO Milano

SIÈGE SOCIAL : 11, rue Leroux - PARIS-XVIe
USINE et SERVICES COMMERCIAUX :
1 et 3, rue S. de Rothschild - SURESNES

I. P. D. - PARIS

La VELAM-ISETTA
n'est pas un modèle réduit de voiture.
Ce n'est pas la réduction d'une voiture
aux dépens de ses occupants,
c'est une voiture nouvelle
qui a été construite « autour de ses passagers » :
elle réussit cette gageure
de posséder, à la fois, des dimensions extérieures
des plus réduites et d'offrir
une habitabilité comparable
à celle des voitures plus importantes.
On peut en juger
par ces quelques chiffres :
— Longueur hors tout ...... 2 m. 38
— Largeur hors tout .... 1 m. 42
— Largeur de la banquette... 1 m. 34

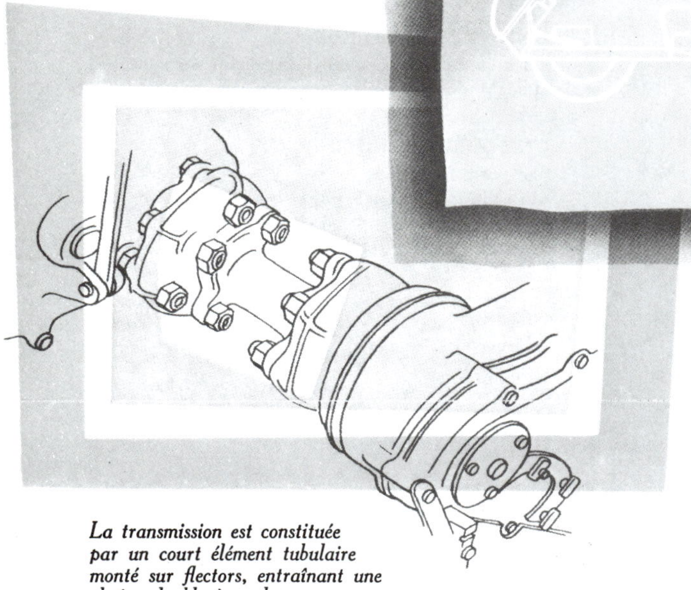

La transmission est constituée par un court élément tubulaire monté sur flectors, entraînant une chaîne double à rouleaux travaillant dans l'huile.

VITESSE : 85 km/h.

CONSOMMATION : 3 l. 5 aux 100 km.

BOITE DE VITESSES : à 4 vitesses et une marche arrière, toutes synchronisées.

VELAM

Isetta

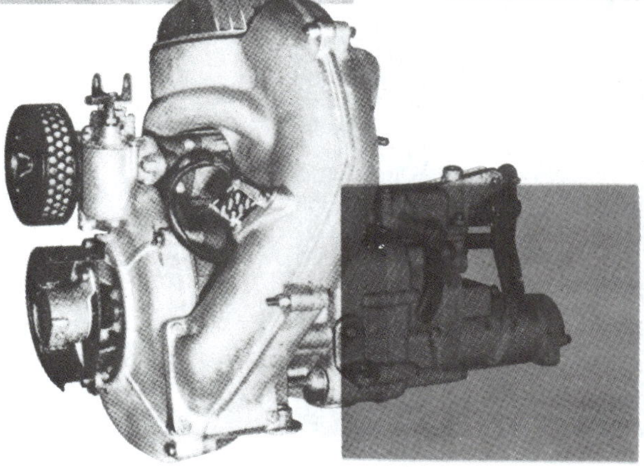

Le moteur de la VELAM-ISETTA *est un 2 temps : deux cylindres en aluminium chromé, chambre de combustion commune.*
Malgré la faiblesse relative de sa cylindrée (236 cm³) qui lui donne une puissance fiscale d'1 CV, ce moteur à très haut rendement et de grande robustesse, développe près de 10 CV à 4.500 t/mn.

La trappe principale d'accès au moteur, située sur la droite de la carrosserie, offre dès son ouverture, la possibilité d'accéder à l'allumeur et au carburateur. Un accès secondaire a été réservé sur la plage arrière de la voiture, juste au-dessus de l'élément moteur et permet la vérification et le démontage éventuel de la bobine et de la bougie.

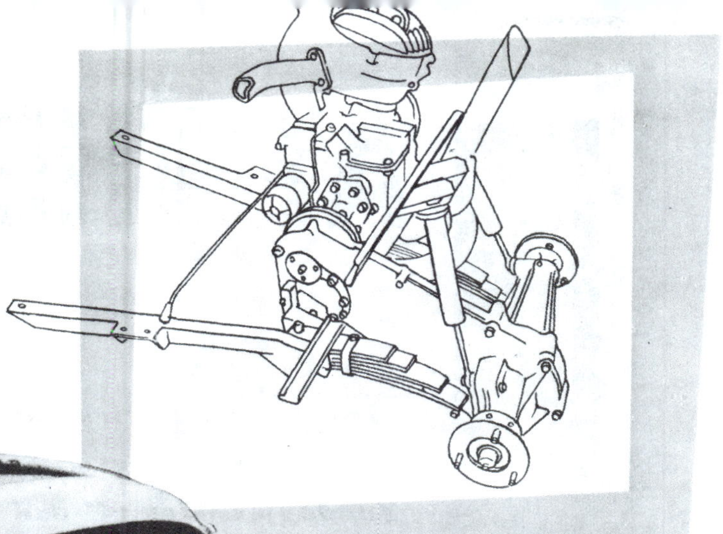

Tous les éléments mécaniques : moteur,
boîtes de vitesses, transmission sont concentrés
sur un faux châssis. Celui-ci est très aisément
démontable en cas de révision ou d'entretien.
Il suffit, en effet, de dévisser les 6 écrous Simmonds
qui le fixent à la caisse
et de désaccoupler les commandes.

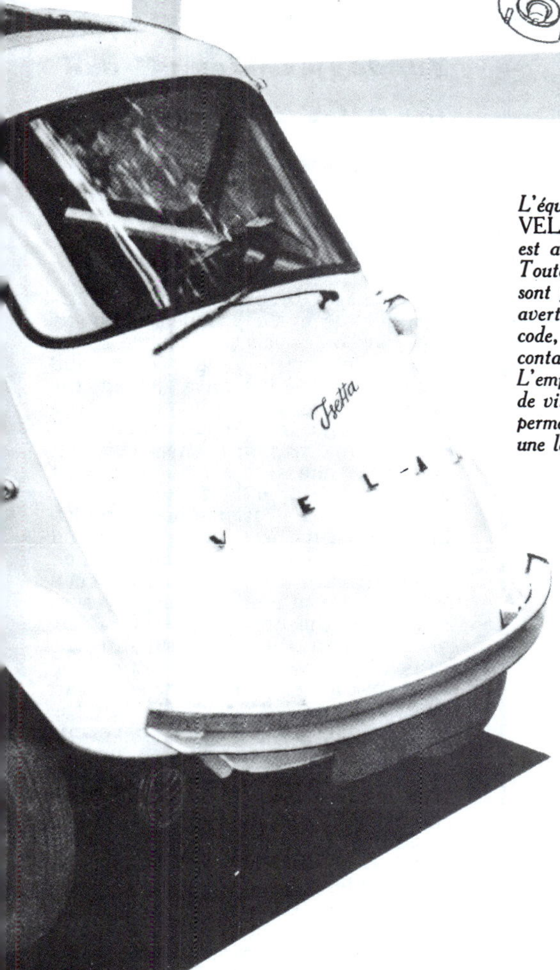

L'équipement de la
VELAM-ISETTA
est absolument complet.
Toutes les commandes électriques
sont groupées au volant :
avertisseur, lanterne, phare,
code, feux de direction à minuterie,
contact, commande du plafonier.
L'emplacement du compteur
de vitesses, situé au centre du volant,
permet au conducteur
une lecture immédiate et rapide.

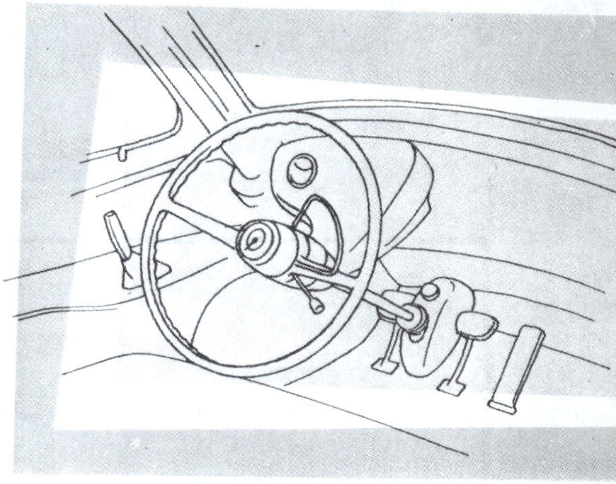

La plus grande originalité de la VELAM-ISETTA
est constituée par la porte frontale qui en permet l'accès :
la banquette est ainsi entièrement dégagée
sur toute sa largeur, la direction s'efface,
ce qui apporte une accessibilité jamais obtenue
jusqu'alors dans aucun véhicule.

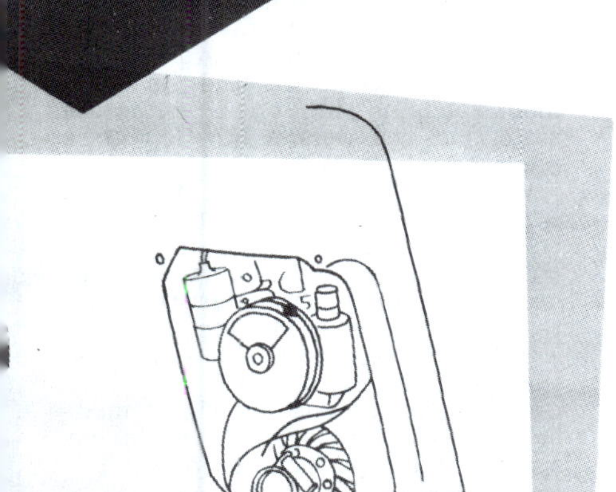

BMW "600" looks like an Isetta from the front, Fiat 600 from rear; has excellent performance and economy.

AT THE

Our European representative tries out two new small cars, Zundapp's "Janus" and the BMW "600"; presents his impressions.

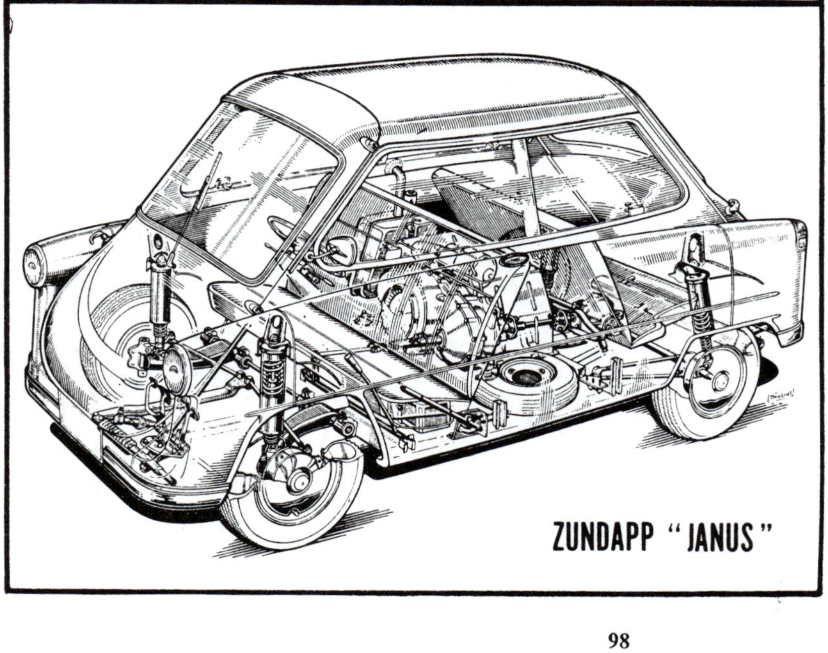

ALREADY bracketing the car market from the smallest economy car to the most expensive sports model, with their Isetta and Type 507, B.M.W. have joined the current move to faster, better engineered and more comfortable economy miniatures with their new 600. It looks like an Isetta at the front and like a Fiat 600 at the rear.

Opening the front door, which brings with it the steering wheel and instrument panel, you walk in and sit down with an ease unapproached by any conventional car of any size. Access to the two rear seats is by one door on the right side of the vehicle. There is really surprising head and leg room at front and rear, and having tried practically all the new crop of baby cars, I have one outstanding impression: the unconventional construction of both the B.M.W. 600 and the back-to-back Zundapp Janus gives advantages in passenger space and ease of entry which cannot be obtained within the same external dimensions on any small car of conventional layout.

Giovanni Michelotti, the young Italian stylist has evolved a body line which. though unconventional, is aesthetically satisfying, and the chassis is such a beautiful study in automobile engineering on a miniature scale that you wonder how it can be made for the money.

Driving the little B.M.W. on a winding road illustrates at once the enormous progress which has been made recently in building the required handling characteristics into rear-engined cars, despite the unavoidable concentration of weight on the rear wheels. The car definitely understeers, and corners with a moderate amount of roll at sports car speeds under perfect control.

The ride is excellent for such a short vehicle, with none of the pitching one might expect. and the engine is as quiet and flexible as are many fours. Brakes pull it up all square, the all-synchromesh gearbox with its short central lever is a joy to use,

. . . from GORDON WILKINS

WHEEL OF BABY

and driving vision, with no hood at all in front, is all one could wish.

The spare wheel is carried in the front door, behind a detachable fibre-glass inner panel, and a heater duct brings warm air from the engine into the door to demist the windscreen. Luggage space is limited, but what can you expect in a full four-seater car which is only 9½ feet long? With only two people on board, there is of course plenty of space on the rear seats.

The torque curve is flat, and there is about 29 lb./ft. available at 2,500 r.p.m., so top gear performance is quite impressive and one can slam the throttle wide open at 15 m.p.h., pulling away with surprising vigour and smoothness.

The front bench-type seat is adjustable, and there is a neat instrument group under the steering wheel consisting of a speedometer, with warning lights for headlamps, traffic signals, and dynamo charge. Levers under the steering wheel control self cancelling indicators, and headlamp flashing. The car goes briskly up to over 50 m.p.h., and the absolute maximum seems to be around 55-60 m.p.h., while fuel consumption is quoted by the factory as 50-65 m.p.g. Imperial.

Janus is unusual looking; has useful performance and very simple controls, including positive-stop gearshift.

Zundapp Janus

Having built some 600.000 motorcycles and scooters, Zundapp were very conscious of the recession in the German market, and some time ago they extended their interests by acquiring the rights in the Dornier Delta prototype, a back-to-back four-seater baby car with front and rear doors.

It has since been redesigned, with hinged doors replacing the original lift-up type, but it retains the same basic layout, and is now in production

The body is a unit structure in steel. to which are attached front wheels on forward-projecting swinging arms, with coil springs and an anti-roll bar. Steering is by rack and pinion, and rear suspension is independent, using a layout similar to that of the Goggomobil. The drive shaft forms one arm of a wishbone, and the other tubular arm projects forwards to a pivot on a cross member under the rear seat.

Suspension media at front and rear are coil spring-damper struts inside the wheel boxes. The engine is a fan-cooled single cylinder two-stroke of 248 c.c. lying on its side in the wedge-shaped space between front and rear seats, which it shares with the spare wheel and fuel tank. It gives 14 b.h.p. at 5,000 r.p.m.

A chain in an oil bath transmits power to a multi-plate clutch also running in oil, and also to a four-speed and reverse non-synchromesh gearbox, from which a short shaft takes the drive to the differential and spiral bevel final drive. Four wheel hydraulic brakes have a total lining area of 75.6 sq. in., and ribs on the face of each brake drum form a fan for cooling.

Seats and backrests fold flat to form a double bed, and with the rear seat only folded, there is quite a big freight space which is easily reached through the rear door. The unusual seating arrangement gives unrestricted head and leg room, and concentrates all the weight within the wheelbase. Riding in the rear is a pleasant relaxing sensation, providing you have confidence in the driver; but what about the dazzle from headlamps of following cars at night?

Zundapp have thought of that, and each rear passenger has a translucent green visor which is adjustable as an eyeshade.

Getting in at front and rear is the easiest thing imaginable. The steering column and instrument panel are fixed, however, and do not come forward with the door, so that the driver must be seated before the front passenger.

On the left is a big lever like a broomstick. which you push to change up and pull to change down, letting it return to the central position after each gearshift. Should you ever be in doubt about which gear is engaged, there is a neat indicator on the instrument panel, with a pointer which reads R.L.O.2.3.4. to show Reverse, First, Neutral, 2nd, 3rd, or top.

The change goes through as fast as you can make it, and is almost foolproof, despite the lack of synchromesh in the gearbox. The brakes seem well up to the job, and the ride is quite surprisingly good, bearing in mind the extremely short wheelbase and the strange seating position. The tiny engine does not at first sight hold out much promise of performance, but with two people on board, the little Zundapp proved a good deal more lively than some other current offerings in the economy car field, and the car seems capable of about 50 m.p.h. flat out. Fuel consumption is said to be about 50 m.p.g. at a steady 40 m.p.h.

ZUNDAPP JANUS SPECIFICATION

Engine, single-cylinder air-cooled two-stroke, 67 x 70 mm., 248 c.c., 6:7:1 compression; 14 b.h.p. at 5,000 r.p.m., max. torque 15.5 lb./ft. at 4,800 r.p.m. Transmission, primary chain and multi-plate clutch in oil. Four speed and reverse gearbox, open propellor shaft, spiral bevel final drive. Overall ratios, 27.97, 14.12, 8.94, 6.74 to 1. Reverse, 26.72 to 1. Suspension: Independent front by swinging arms, coil springs and ati roll bar. Independent rear by swing axles with radius arms and coil springs. Telescopic dampers.

Dimensions: Wheelbase, 71.8 in. Track front, 43.5 in.; rear, 43.6 in. Length, 110.6 in. Width, 55.5 in. Height, 55 in. Weight, 935 lb. ●

Minicars, an entirely new concept in automobiles has sprung up in the midst of the economy car boom. The two or four passenger vehicles are midway between the small economy cars and the two-wheeled motorscooters. Most feature conversions of well proven continental cycle engines.

minicars

Gas mileage borders on the fantastic while performance surprises big car drivers. The distinctive scootabouts are perfect for city driving where traffic is heavy and parking spaces light. Their low price tags and low maintenance costs make them a worthy second car or excellent first car purchase for families with limited budgets.

BERKELEY

The racy Berkeley is the world's smallest sports car. The 322 cc. baby-sized engine will push the fiberglass-bodied roadster to a top speed of 70. The price tag is roughly $1,000 depending on extras.

VESPA

Vespa is built in France from Italian design. It features a rear-mounted air-cooled engine, and independent four wheel suspension.

FRISKY

This $700 baby, the Frisky, features gull-wing doors a la the Mercedes-Benz.

The Frisky becomes a convertible by removing the doors. Two adults and three children can travel comfortably in the car. Gas mileage is a phenomenal 60 miles per gallon. The model arrives here soon.

GOGGOMOBIL

The Goggomobil gives the best perormance of the minicars in addition to being the prettiest.

The GLAS parallel-twin engine of the Goggomobil develops 17 horsepower on only 293 cubic centimeters. The four speed gear box is pre-selective making driving a cinch. The engine, gearbox and transmission are combined in a single unit. Swing axles give it true sports car handling.

HEINKEL

The Heinkel Cabin Cruiser is a three wheeler which is fast, roomy and stylish.

The '58 Heinkel Cabin Cruiser has a one cylinder engine which puts out 9½ horses. Independent suspension helps handling.

ISETTA

BMW's four-wheeled Isetta will park in the smallest spaces. It's the perfect car for beating traffic jams. Low operating costs make it an attractive buy.

The Isetta uses a powerful four-cycle overhead valve engine. A heater and bumper guards are standard equipment.

BMW

ISETTA 600

Unusual two-door miniature sedan offers comfort for five; up to 58 mpg economy

● The success of the BMW Isetta 300, a rear-engined, two-passenger, single-door miniature car, has prompted the manufacturers to develop a somewhat larger and more versatile model. The new car, the Isetta 600, is a four-five passenger vehicle built on a 67" wheelbase, 11" shorter than the Fiat 600 and 27" shorter than the VW. Although the 1,135-pound car has an over-all length of only 115", interior roominess compares very favorably with its larger competitors.

One of the most striking features of the new model is (like the 300) the forward-opening door which gives easy access to the front seat. The steering column is double-jointed to permit this. The 600 has a second door, on the right side of the lengthened body shell, for access to the rear seat. In keeping with the body modifications, the rear tread was widened from 20" (on the 300) to 25⅔".

Powered by a horizontally-opposed, two-cylinder, air-cooled, 26-hp engine mounted in the rear, the 600 has a claimed top speed of 68 mph and fuel economy of up to 58 mpg. Suggested retail price is $1,398 f.o.b. P.O.E.

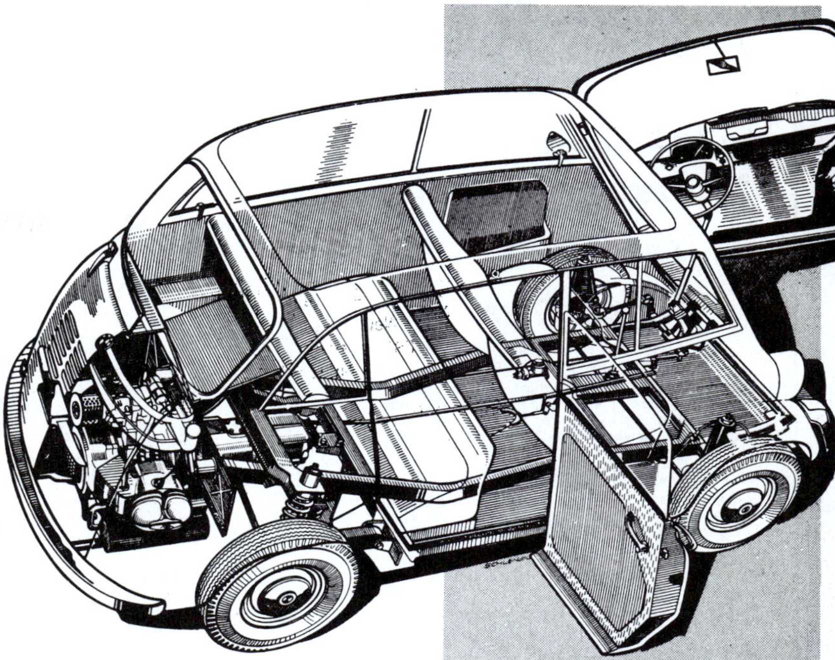

Isetta 600 has an ultra-short (67") wheelbase—only 9" longer than the 300—yet carries five in comfort. The suspension is independent all-around with coil springs front and rear. Rear tread is wider than in 300 model.

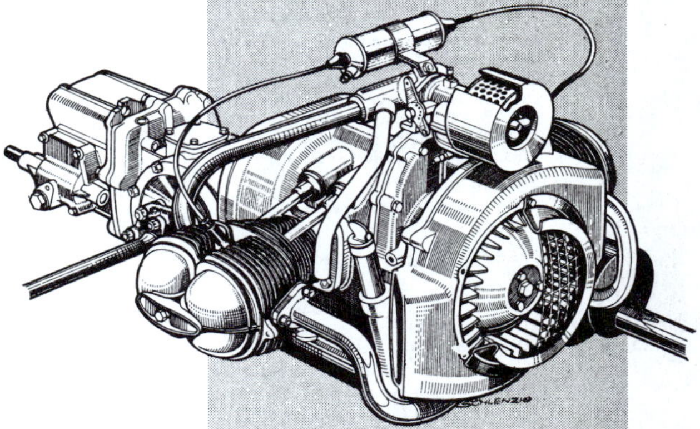

Transmission, mounted forward of two-cylinder, air-cooled engine, has all four forward speeds synchronized for easy shifting. Air ducts to route air around finned cylinders are not shown here. The two-door styling is startling.

The Motor Road Test No. 30/58

Make: Isetta **Type:** "Plus" 3-wheeler

Makers: Isetta of Great Britain Ltd., New England Road, Brighton, Sussex

Test Data

World copyright reserved; no unauthorized reproduction in whole or in part.

CONDITIONS: *Weather: Mild and showery with moderate breeze (Temperature 62°-65° F., Barometer 29.8 in. Hg.). Surface: Smooth tarred macadam. Fuel: Standard-grade pump petrol (approx. 80 Research Method Octane Rating).*

INSTRUMENTS

Speedometer at 30 m.p.h.	2% slow
Speedometer at 50 m.p.h.	2% slow
Distance recorder	2% slow

WEIGHT

Kerb weight (unladen, but with oil and fuel for approx. 50 miles)	7 cwt.
Front/rear distribution of kerb weight	55/45
Weight laden as tested	10 cwt.

MAXIMUM SPEEDS

"Maximile" Speed (Timed quarter mile after one mile accelerating from rest.)

Mean of four opposite runs	51.9 m.p.h.
Best one-way time equals	53.9 m.p.h.

Speed in gears

Max. speed in 3rd gear	41 m.p.h.
Max. speed in 2nd gear	28 m.p.h.

FUEL CONSUMPTION

88.5 m.p.g. at constant 20 m.p.h. on level.
83.5 m.p.g. at constant 30 m.p.h. on level.
71.5 m.p.g. at constant 40 m.p.h. on level.
48.0 m.p.g. at constant 50 m.p.h. on level.

Overall Fuel Consumption for 582 miles, 10.6 gallons, equals 54.9 m.p.g. (5.15 litres/100 km.).

Touring Fuel Consumption (m.p.g. at steady speed midway between 30 m.p.h. and maximum, less 5% allowance for acceleration), 66 m.p.g.

Fuel Tank Capacity (maker's figure) 2.8 gallons (including 0.65 gallon reserve).

STEERING

Turning circle between kerbs:

Left	27¾ feet
Right	27 feet
Turns of steering wheel from lock to lock	2¼

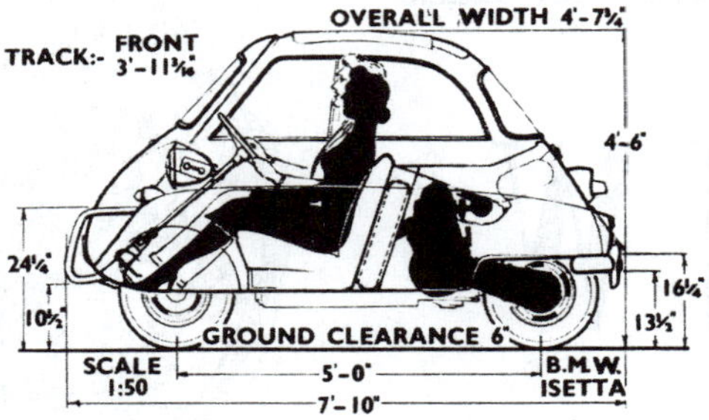

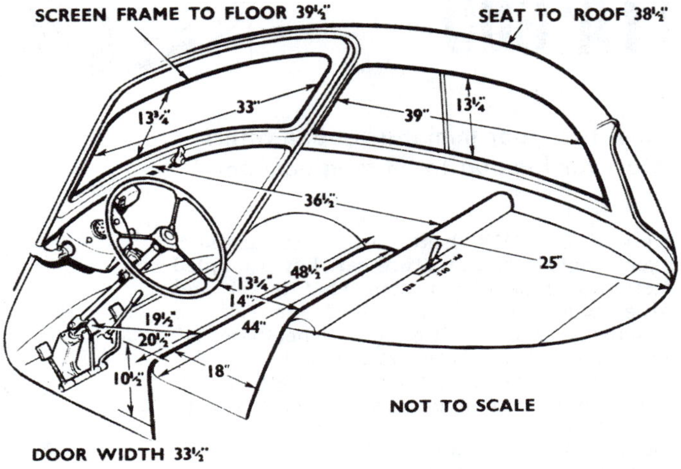

ACCELERATION TIMES from standstill

0-30 m.p.h.	11.3 sec.
0-40 m.p.h.	19.4 sec.
0-50 m.p.h.	41.5 sec.
Standing quarter mile	27.9 sec.

ACCELERATION TIMES on Upper Ratios

	Top gear	3rd gear
10-30 m.p.h.	20.5 sec.	12.1 sec.
20-40 m.p.h.	23.1 sec.	16.1 sec.
30-50 m.p.h.	38.4 sec.	—

BRAKES from 30 m.p.h.

1.00g retardation (equivalent to 30 ft. stopping distance) with 80 lb. pedal pressure
0.87g retardation (equivalent to 34½ ft. stopping distance) with 50 lb. pedal pressure
0.38g retardation (equivalent to 79 ft. stopping distance) with 25 lb. pedal pressure

HILL CLIMBING at sustained steady speeds

Max. gradient on top gear 1 in 16.6 (Tapley 135 lb./ton)
Max. gradient on 3rd gear 1 in 11.1 (Tapley 200 lb./ton)
Max. gradient on 2nd gear 1 in 7.3 (Tapley 305 lb./ton)

Maintenance

Sump	3 pints, S.A.E. 40 summer, S.A.E. 20 winter
Gearbox	1 pint, S.A.E. 40
Rear axle	½ pint, S.A.E. 40
Steering gear lubricant	Grease
Cooling system capacity	nil (air cooled)
Chassis lubrication	By grease gun every 1,000 miles to 4 points
Ignition timing	8° b.t.d.c. static
Contact-breaker gap	0.016 in.
Sparking plug type	Lodge HH14
Sparking plug gap	0.024 in.
Valve timing	Inlet opens 6° a.t.d.c. and closes 34° a.b.d.c. Exhaust opens 34° b.b.d.c. and closes 6° b.t.d.c.
Tappet clearances (cold)	Inlet 0.004—0.006 in. Exhaust 0.006—0.008 in.
Front wheel toe-in	3/32 to 1/16 in.
Camber angle	1½°
Castor angle	5°
Steering swivel pin inclination	12°
Tyre pressures	Front, 16 lb. Rear, 24 lb.
Brake fluid	Girling crimson
Battery	12 volt, 32 amp. hr.
Miscellaneous	Final drive chain tension adjustable

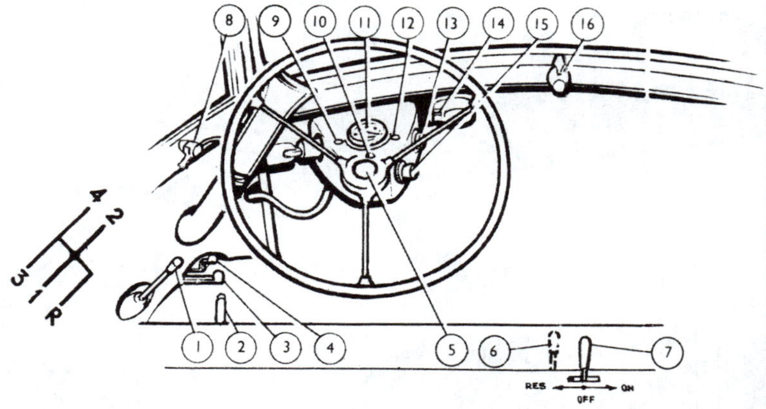

1, Gear lever. 2, Handbrake. 3, Heater control. 4, Choke control. 5, Horn button. 6, Demister control. 7, Petrol tap. 8, Headlamp dip switch. 9, Direction indicator warning light. 10, Headlamp main beam indicator lamp. 11, Speedometer and distance recorder. 12, Dynamo charge warning light. 13, Ignition and starter switch. 14, Windscreen wiper motor and switch. 15, Lights switch. 16, Direction indicator switch.

The Isetta 300

COMPACT in overall length, even with the unusual tubular bumpers used to guard the front door, the Isetta is neat in appearance and easy to park. Entry to the coupé body is from the front, the bench seat having room for an adult and two children; the steering column swings forward with the door to simplify access to the driving seat.

Three-wheeled Version of a Well-tried Miniature Convertible

CONVERSION of the Isetta 300 coupé from four to three wheels, by elimination of the former narrow-track solid rear axle, is a change which in Britain leads to a diminution in the rate of purchase tax levied and also to a reduction in the annual licence fee from £12 10s. to £5. Whilst certain insurance companies may still treat three-wheeled cars with exaggerated caution, we have been able to verify that the Isetta made into a tricycle is fully as serviceable a miniature car as it is already known to be in four-wheeled form. The main limitation is that, in order that the weight of an unaccompanied driver shall balance that of the engine, left-hand drive only is offered, an arrangement which scarcely proves objectionable either in city traffic or in open country, but which does interfere with visibility when pulling out to overtake other cars on busy out-of-town roads.

In respect of road manners, the Isetta three-wheeler has lost little of the merit of the four-wheeler. On the open road at speeds in the 40-50 m.p.h. region there is a slight loss of straight-running stability, as if the oversize rear tyre has slight sideways flexibility on a too-narrow rim, this effect being more evident two-up than when the driver is alone, but soon ceasing to be noticed. Experiments with really tight turns on an adhesive concrete surface showed that by fierce cornering it was possible to lift a front wheel off the ground, but this was only achieved two-up and turning to the left, only by really deliberate effort. Extremely effective braking with quite modest pedal pressures will be noted on the data page, although whereas normal stops were made in a straight line, fierce braking could induce some snaking; the pull-up lever hand-brake used on "Plus" models works well, although its location below the driver's legs is not very convenient. The springing allows some roll on corners but absorbs road shocks surprisingly well.

Cruising at 45 m.p.h.

The performance figures for this 300 c.c. car, showing a top speed comfortably in excess of 50 m.p.h., are highly creditable and better than those of the heavier four-wheeled Isetta tested by us in 1957. As is our usual practice, they were recorded with two men in the Isetta, which when driven solo becomes very much more lively. Almost new and rather stiff when handed over to us, the four-speed gearbox soon began to show its true merit, it being normal to run up to nearly 40 m.p.h. in third gear and to cruise at 45 m.p.h. on the open road. The very low first gear is at times a nuisance, but will easily re-start the car on astonishingly steep hills. Whilst the need to make free use of the four gears (which have a "back-to-front" gate arrangement) does not make this an especially easy car to drive, a competent man or woman soon becomes thoroughly at home in it. Starting the engine from cold requires some correlation of choke and accelerator, but can be quick and easy.

The electric starter which is also the dynamo is completely silent in operation, and is controlled by turning the ignition key beyond the "on" position. Absence of any water jackets or radiator results in the air-cooled engine being quite quick to warm up after being started or to cool down when standing, so far as response to the throttle is concerned. Full power is not, however, available until the engine and transmission lubricants have also had time to attain their running temperatures.

Noise from the hard-working little air-cooled engine is well silenced, apart from some rattle from the rubber-suspended silencer at low r.p.m., but whine from the transmission or cooling fan is rather prominent at main road cruising speeds.

In Brief

Price £291 1s. 6d. as tested, plus purchase tax £74 18s. 0d. equals £365 19s. 6d.

Price without "Plus" equipment (including purchase tax) £339 19s. 6d.

Capacity	298 c.c.
Unladen kerb weight ...	7 cwt.
Acceleration:	
20-40 m.p.h. in top gear ...	23.1 sec.
0-50 m.p.h. through gears	41.5 sec.
Maximum direct top gear gradient	1 in 16.6
"Maximile" speed	51.9 m.p.h.
Touring fuel consumption ...	66 m.p.g.

Gearing: 9.4 m.p.h. in top gear at 1,000 r.p.m.; 19.6 m.p.h. at 1,000 ft./min. piston speed.

FAN-COOLED, the 298 c.c. single-cylinder o.h.v. engine of B.M.W. manufacture is mounted ahead of the rear wheel, on the right of the car. A separate panel gives access to the sparking plug from above.

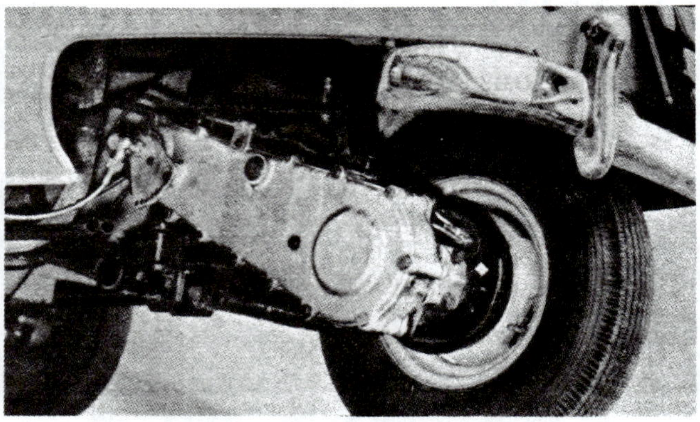

CONVERSION of the Isetta to a single rear wheel allows the adjustable fully-enclosed rear chain to be retained.

The Isetta 300

Essentially the Isetta is a two-seater, but its rather hard bench seat is wide enough to make an adult and two fair-sized children at ease. The forward door soon proves itself to be entirely practical, but tall drivers were apt to find their left knee inconvenienced by a projection from the steering column and also by the gear lever when reversing—if reverse gear is rendered inoperative, this runabout can legally be driven by a 16-year-old with a motorcycle licence, but without reverse gear the 27-28 ft. turning circle would prove insufficiently compact for convenience.

Practical amenities include rather stiff-to-move sliding glass side windows and an excellent little folding roof panel for fine days (or for fine minutes, so quick is the process of opening and closing the roof) plus a heater system which on a cool night in August de-misted the screen effectively, but seemed rather limited in heat output. Minor luggage can go on a huge shelf behind the seat, and there is a map pocket, a single-blade electric wiper for the screen (which is flanked by unpleasantly wide pillars), flashing turn indicators, excellent dipping headlights, and a petrol reserve tap, but no fuel contents gauge.

In respect of petrol economy, the Isetta runs happily on the cheapest grades of fuel, but its consumption rate varies sharply with driving technique—as witness steady-speed figures which on a scarcely run-in car ranged from 88½ m.p.g. at a mere 20 m.p.h. to 48 m.p.g. at a bustling 50 m.p.h. As it seems natural either to use the Isetta as a runabout for frequent short runs which hardly warm up the engine and transmission, or on longer journeys to make much use of the gears for acceleration, it is optimistic to expect much more than 50-60 m.p.g. As this is a four-stroke engine with pressure lubrication from a three-pint sump, only slight oil consumption should have to be allowed for in predicting penny-a-mile petrol and lubricant bills.

For many people, the virtue of the Isetta which will be even more important than its low cost and inexpensive thirst for petrol is its compactness. With an overall length including bumpers of only 7 ft. 10 in. it can be garaged in a very small area, and parking places are exceptionally easy to find; apart from the possibility of nose-to-kerb parking, it is possible to go forwards into a short gap between other parked cars and stop at an angle to the kerb without needing to reverse into position, the narrow tapering tail of the Isetta not projecting obstructively into the roadway.

Perhaps primarily as a "second car" but also as a one-man business runabout or a vehicle for a small family, the Brighton-built Isetta in this three-wheeled form gains in attractiveness from its reduced annual licence fee. Within the visible limits of its carrying capacity it is, in the old phrase, willing to go anywhere and do anything, and the manner of its going entertains some people very much indeed.

Mechanical Specification

Engine

Cylinders	1 (fan cooled)
Bore	72 mm.
Stroke	73 mm.
Cubic capacity	298 c.c.
Piston area	6.3 sq. in.
Valves	Pushrod o.h.v.
Compression ratio	7/1
Carburetter	Bing 1/24
Fuel pump	Gravity feed
Ignition timing control	Centrifugal
Oil filter	Full flow
Max. power (net)	13 b.h.p.
at	5,200 r.p.m.
Piston speed at max. b.h.p.	2,490 ft./min.

Transmission

Clutch	Single dry plate (Fichtel & Sachs)
Top gear	6.1
3rd gear	8.17
2nd gear	12.18
1st gear	23.21
Reverse	30.0
Final drive	Enclosed chain
Top gear m.p.h. at 1,000 r.p.m.	9.4
Top gear m.p.h. at 1,000 ft./min. piston speed	19.6

Chassis

Brakes	Hydraulic on 3 wheels
Brake drum internal diameter	7.1 in.
Friction lining area	50 sq. in.
Suspension:	
Front	Dubonnet-type i.f.s. by coil springs
Rear	¼-elliptic leaf springs
Shock absorbers	Girling telescopic
Steering gear	B.M.W. worm and nut
Tyres:	
Front	4.80—10
Rear	5.20—10

Coachwork and Equipment

Starting handle ... None
Battery mounting ... Under seat
Jack ... Screw-type
Jacking points ... Front cross member and rear wheel carrier
Standard tool kit: Tool roll containing 5 open-ended spanners, sparking plug spanner and tommy bar, screwdriver.
Exterior lights: 2 headlamps, 2 sidelamps, 2 tail lamps, 1 number plate/stop lamp.
Number of electrical fuses ... 1
Direction indicators: Amber flashers (non self-cancelling) on sides of body.
Windscreen wipers ... Single-blade electrical
Windscreen washers ... None

Sun vizors ... None
Instruments: Speedometer with non-decimal total distance recorder.
Warning lights: Direction indicators, head-lamp main beam, dynamo charge.
Locks:
With ignition key ... Ignition/starter switch
With other keys ... Front door, engine cover
Glove lockers ... None
Map pockets ... One inside front door
Parcel shelves ... One behind seat
Ashtrays ... None
Cigar lighters ... None
Interior lights: Instrument lighting only (with front door open, right sidelamp illuminates body interior).

Interior heater: Fan-circulated hot air from engine and exhaust system ducted to body interior and windscreen.
Car radio ... Optional extra
Extras available: Radio, luggage carrier, locking petrol cap, inspection lamp, seat covers, wheel embellishers, spare wheel, two-colour finish, windscreen washers, interior grab handles, etc.
Upholstery material ... Cloth
Floor covering ... Rubber mats
Exterior colours standardized ... 4 (two-tone at extra cost)
Alternative body styles ... None

BMW 600

Although designed primarily to pack exceptional carrying capacity within modest external measurements, the BMW 600 has undeniable poise, and there is nothing freakish about it

WHEN the Bavarian motor company set about designing the BMW 600, they began with two important ready-made assets. One was a wealth of knowledge and experience gained in building the little single-cylinder Isetta, the so-called bubble car: the other was a full production line already turning out the famous R.60 flat-twin, air-cooled engine, which some consider to be the world's finest motor-cycle power unit. Out of these has grown a real little four-seater motor car with grown-up manners, a lively performance and low running costs.

Like the Isetta, the 600 has a separate, tubular frame and independent front suspension by an unusual system; this has a Dubonnet-like leading link arrangement (whereby the steering mechanism is isolated from vertical road wheel movements) in conjunction with co-axial coil spring and damper units. At the rear it has a more ambitious layout than the Isetta—a differential, a track only 2.3in less than that at the front, and each wheel independently mounted to a rigid box-section trailing arm having widely spaced pivot bearings. Each suspension unit consists of a co-axial coil spring and telescopic damper, supplemented by rubber buffers giving a progressive build-up of reaction. In conjunction with the 582 c.c. engine (rear mounted) are a conventional car-type single-plate clutch and a gear box with four forward speeds, all with baulk-ring type synchromesh.

A full four-seater body with very compact external dimensions retains an Isetta feature, the front end-door. Rear passengers enter through a side door on the right—better suited to those countries where the rule-of-the-road enables them to alight directly on to the pavement rather than into the traffic stream.

Although the forward door is novel, it is not wholly practical. On the car tested, it was self-supporting when opened only if the car was standing on a camber favourable to it; ajar in wet weather, it allows rain to fall directly on the front occupants' knees. Whereas one soon learns to avoid parking too close to a car ahead, one cannot prevent another driver from parking his car, in one's absence, so close to the 600 that this door cannot be opened sufficiently to enter. A technique has to be developed for getting in and out easily and—for women—gracefully; it is particularly tricky for the driver if the passenger remains seated.

Certainly this door is substantial, strongly hinged and efficiently sealed against the weather. It has a particularly well-engineered locking claw. The telescopic and universally-jointed, three-piece steering shaft, allowing the wheel to be bracketed to the door, had no lost motion after 16,000 miles recorded on the car tested. The driver's electrical controls also move with the door, their cables carried to the body through an armoured, flexible conduit.

These controls include a push-pull lamp switch, a key-switch starter and a wiper switch on the instrument panel. Behind the two-spoke steering wheel are two finger-tip levers, that on the left for head lamp flash and dip, that on the right for direction signals and horn. The solitary instrument dial is a speedometer with total distance recorder, and there are coloured lamps to indicate head lamp main beam (blue), generator charge (red) and flashing turn signals (green). The vertical gear lever and powerful, pull-up hand brake are to the driver's right in this left-hand-drive vehicle. To his left is a group of three horizontal levers controlling, from top to bottom, engine starting mixture, interior heating and the fuel supply—main, off, and reserve. This reserve supply constitutes about two-thirds of a gallon in a total capacity fractionally over five.

The front seat is of the bench type with a fixed back-rest, shaped to give restful support in normal driving as well as during fast cornering. Trim is in a hard-wearing synthetic cloth, in conjunction with a smooth plastic material. The seat cushion is resilient enough to conform to the occupants' shape, but not so soft as to cause much bouncing within the car: in other words, their insulation from road shocks is governed almost entirely by the chassis suspension. The seat is adjustable fore and aft, locked by wing nuts, but drivers of average height found that they did not need

Front occupants enter through this end door. The steering shaft is telescopic and universally jointed to allow the steering wheel to move with it. There are side armrests on the lips of the map pockets

to have it at its fully back position, even with no rear passengers' knees to consider.

All controls are ideally placed. Clutch and brake pedals are at either side of the steering shaft; brake and accelerator pedals can be operated simultaneously with the right foot for downward changes on approaching, say, a roundabout or sharp corner. The two screen-wiper arm pivots are placed off-centre, to give the driver the most comprehensive forward view in wet weather, and the wiper motor, which is self-parking, runs quietly. An interior mirror, providing adequate rear view without obscuring forward vision, is complemented by an exterior mirror on the left. The screen pillars are fairly thick, but the driver is so near to the front of this vehicle that their obstruction is not serious. As stated already, there is only one side door—on the right—so that the car's two flanks are different. On the left the driver has a long sliding window, which opens sufficiently for him to put his head out if necessary when reversing, and the left rear passenger also has a slide-forward side window. Both these panes can be locked whilst fractionally open to allow some ventilation when the car is parked. On the right there is a short, sliding pane at the front, but the frameless door window and quarter-light behind it are fixed. For maps and oddments there are capacious pockets in the body walls at either side of the front seat, a pad being provided on the lip of each of these.

The rear bench seat is acceptably comfortable, and even with four aboard the BMW feels larger inside than it looks from the outside. Beneath this seat are housed the tool kit, jack and battery, leaving sufficient space to stow other small oddments there. With the rear seat in use, there is rather limited baggage space behind its squab, plus a little room for oddments under the front seat. When only two are travelling, the back squab can be folded forward to carry much more luggage. Squab and cushion are easily removed to add further to the loading volume. A wool cloth lines the interior of the car above the waist, except for the centre section of the roof, which is trimmed with plastic over a synthetic foam lining; this has sound-deadening characteristics as well as adding something to passenger safety.

Naturally, the relatively small air volume within the BMW 600 leaves it prone to misting in cold weather, especially whilst carrying a full complement of passengers. There is, however, in-built provision for heating and demisting. This hot-air supply is ducted from the engine cooling baffles to independent adjustable valves by the front and rear passengers' feet, and to a fan-shaped outlet by the left front pillar, aimed to blow over the screen. An extra supply to keep the rear window clear would be appreciated.

When starting the engine from cold the throttle pedal must be left alone, since the choke is interconnected with the carburettor butterfly to speed up its idling setting. Also an accelerator pump is fitted, so that pumping of the throttle pedal results in an over-rich mixture and difficult starting. Once the correct technique has been acquired, an initial difficulty over starting is overcome; as with all air-cooled engines, the cylinder temperature drops quickly during even comparatively brief stops, so that the choke control must be used more frequently than with a liquid-cooled engine.

Any tendency to stall whilst cold is dealt with quickly by a jab on the accelerator to work the accelerator pump, but the engine warms very quickly and soon pulls strongly. The tested car's cold tick-over was somewhat erratic, but reasonably even when warm. There is an astonishing lack of vibration throughout the range, due largely to really effective rubber engine and transmission mountings—clearly the result of much research. It is interesting that the "floor" of the engine compartment is attached to the power unit itself, to permit movement without affecing the body structure.

For the throttle pedal there is a two-pressure return spring, the second pressure presumably to discourage uneconomical driving methods. The first pressure limit provides reduced acceleration, but almost full maximum speed under most conditions. This second pressure seemed unnecessarily strong; many drivers probably would prefer to do without it and judge throttle openings for themselves.

Beneath a five-gallon fuel tank, the flat-twin engine is neatly installed, with comprehensive ducting from the crankshaft-driven fan to cool its cylinders. Individual 6-volt coils are wired in series from a 12-volt electrical system

Fuel economy is especially important with a vehicle of this type. During its road test, the car under review covered many miles in dense city traffic as well as being driven hard for much of its country mileage. Although the overall average consumption quoted is in the forties, most A to B journeys were covered at appreciably better than 50 m.p.g.; more considerate use of its powers brought about figures in the middle fifties.

Never obtrusive to an annoying extent, the engine is very commendably quiet on light throttle openings at any speed. Naturally, it is heard and felt most by the rear passengers, who sit very close to it.

While the R.60 engine is made with a power output of 35 b.h.p. for motor cycles, the related version used in the 600 car is restrained to a mere 19.5 b.h.p., in view of German insurance rates, which are much more sympathetic financially towards cars having under 20 b.h.p. This relatively low power output from an efficient engine should result in long life and freedom from minor troubles, while the restriction of its induction passages gives it particularly good low-speed torque characteristics. Although the makers claim that the BMW can be brought down to 15 m.p.h. in top gear, on the test car it was found that about 20 m.p.h. was the practical low limit.

In a car of limited power, frequent use needs to be made of the gear box, and this unit, having very effective synchromesh on all four forward speeds and a light and precise change, encourages frequent use of the indirects. Such a practice is recommended, because neither engine cooling nor dynamo charge are very powerful at low crankshaft speeds. Not only is this box a joy to use, but

Perhaps the BMW 600's most attractive angle of view, this shows the single side door and the wide curved rear window. The engine compartment is ventilated by a minimum of louvres

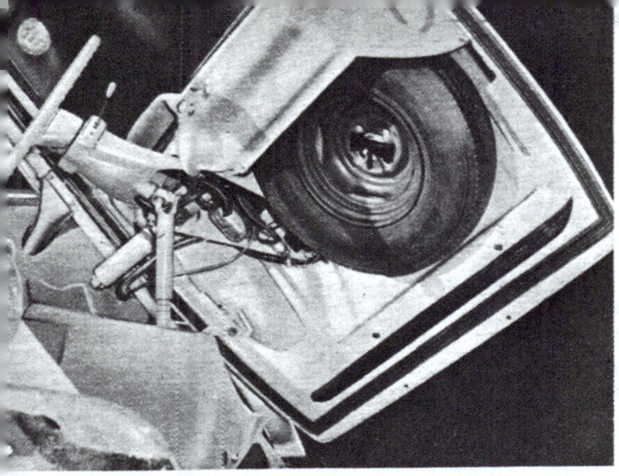

Left: Normally concealed behind a moulded, plastic panel, the spare wheel is bolted to the front door. The screen heater duct can be seen beneath the steering wheel. Right: There is no door on the left side, which has two long, sliding windows. An external mirror is a standard fitting

BMW 600 . . .

its ratio spacings seem ideally suited to the nature of the car and its engine characteristics. In first, second and third speeds the makers recommend everyday maxima of 16, 28 and 44 m.p.h., but the engine shows no distress if extended to the maxima quoted in the performance panel—that is, 20, 36 and 52 m.p.h.

In top, the cruising speed recommended is the same as the maximum, the final drive ratio being calculated to prevent overspeeding of the engine on its present power output. Fourth in the gear box is, in fact, an indirect overdrive (0.86), so that, in conjunction with final drive gears giving a 5.43 step-down, the overall top ratio is 4.6 to 1. This sounds relatively high, but the small wheels result in only 12.4 m.p.h. per 1,000 r.p.m.—or 5,000 at 62 m.p.h., the appropriate maximum and top cruising speed.

Able to reach 50 m.p.h. from a standing start in 22.4sec, the BMW can more than hold its own in an average main road traffic stream, while its compact dimensions and stability on the road enable an alert driver to thread it very quickly indeed through suburban or city traffic. On the open road, it seemed happiest at between 50 and 55 m.p.h., which it maintained indefinitely where conditions permitted. Its hill-climbing properties and low-speed pulling power are surprisingly good, but the relatively high third gear, which can deal with gradients up to 1 in 9, ensures that a commendable average speed can still be maintained in hilly country. With four up, the little engine is naturally handicapped more than would be one of greater capacity.

When taking to the BMW after driving larger cars, one notices at first some thudding of the little wheels, which make heavier work of negotiating small potholes and rough road surfaces than would larger ones. The front occupants are subjected to a certain amount of low-frequency up-and-down movement over such roads, being so nearly above the front wheels. To the rear occupants, nearer to the middle of the vehicle, this movement would be more correctly described as pitching. It is, however, never excessive, and the general comfort of the car is very good. At higher speeds the ride becomes much more level, and the general feel is of a much larger vehicle. Rather unexpected, in view of the small wheels and the short wheelbase of the car, is its ability to travel quite fast over considerable humps or waves in the road; the passengers are then not thrown about much, nor do the dampers allow any such excess movement to be continued. Certain road surfaces of coarse texture excite some rumble and resonance through the structure.

Cornering abilities of the BMW are exceptional. It rolls very little, and in normal use there is no tyre squeal. The makers recommend appreciably lower pressures for the front tyres than for the rear (17 and 22 lb sq in respectively when two occupants are carried), and with these pressures there is a slight understeer characteristic with two up; thus there is never any question of the tail of the car getting out of hand. The 600 can be flicked left-right-left through an open roundabout with great accuracy and stability.

With a relatively light load its steering properties are admirable, but when fully laden it needs a little more attention at the wheel. The steering is no more than reasonably light at manœuvring speeds, but is always precise and completely free from road shocks. This is due in part to the front suspension system already mentioned, and in part to the fitting of a damper to the steering linkage. For the size of car, the turning circle is not compact; the car tested would turn more sharply to the right than to the left, which favours it for U-turns in wide streets in this country.

Complementing these other handling qualities is a first-class hydraulic braking system. The braking effort is divided well between the four wheels, because even under maximum retardation there was no tendency for the front or rear wheels to lock independently. This feature is of particular value on a wet or otherwise slippery surface, when more caution must be taken to ensure that one is not rammed from behind than to avoid hitting a car ahead. The hand brake, applied to the back wheels, has sufficient power to be used on the move, should any trouble develop with the main braking system. It seemed powerful enough to hold the car securely on any gradient on which the tyres would grip. The BMW was found reluctant to restart from a one-in-three gradient with only the driver aboard, but it made a confident getaway on one-in-four.

This little 600 could be driven at night as fast as traffic and road conditions would allow, having a main lighting system which calls only for praise. The beams are powerful and really well diffused, so that one has full visibility for fast driving along winding roads. The head lamps have asymmetrical dipped beams in the latest continental fashion, with an extra "feed" of light along the kerb for picking out pedestrians and cyclists whilst faced with oncoming traffic.

Left: With the front seat placed for an average driver, there is adequate space for rear passengers' legs. The door opening is wide and deep, and there is plenty of headroom. Limited luggage space is provided behind the squab. With this hinged forward (right) luggage capacity is much increased

BMW 600 . . .

Although no reversing lamp is fitted, the stop-lamp bulbs give plenty of illumination if the brake pedal is applied lightly in conjunction with the accelerator. Alternative illumination for reversing, which obviously should not be used whilst other traffic is about, is provided by selecting one of the direction signals. An interior lamp is fitted for the convenience of passengers; without this switched on, the front compartment is indirectly illuminated by reflection from the lamps off the door when it is opened.

Maintenance of the BMW is simple and straightforward. There are ten greasing points calling for attention about every 1,200 miles, and even such a major job as removing the engine is said to occupy minutes rather than hours. BMW engines are noted for oil tightness, and the engine compartment of this car was no exception. As with all rear-engined cars, it keeps almost free from road dust and water; the layout of this installation is extremely neat, with plenty of space around the main components. It is worth noting, too, that the clean shape of the car makes it easy to wash, while the mudguarding is effective in keeping its flanks clean during rough weather.

In sum, the BMW 600 is a thoroughly practical, economical little vehicle built to the high engineering standards expected of its makers. It has real character—indeed, its owner might come to regard it as something slightly more personal than just utility transport. In its native Germany, free from purchase tax and import duty, it is cheap to buy; but in the U.K. it has to meet very strong competition from our own small cars. Despite this handicap, the BMW has sufficient merit to entitle it to serious consideration.

BMW 600

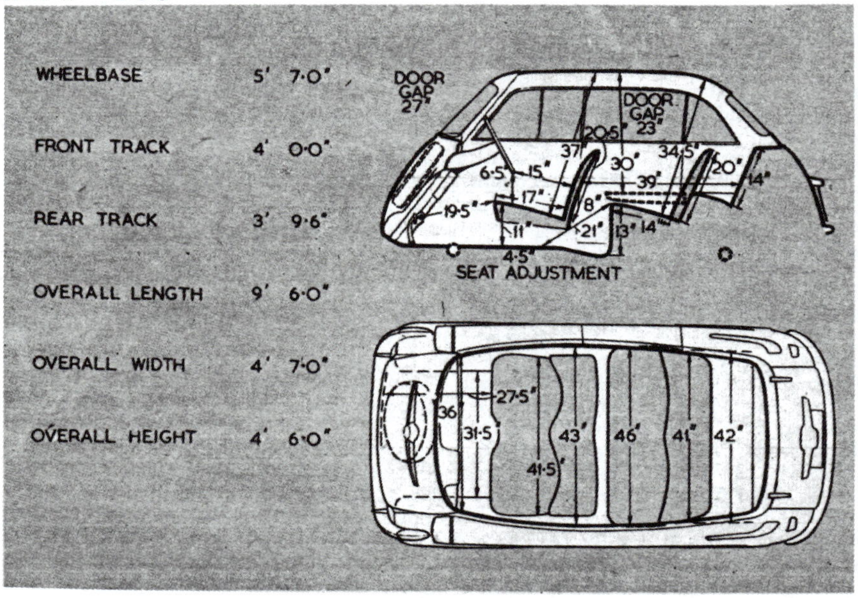

Scale ¼in to 1ft. Driving seat in central position. Cushions uncompressed.

PERFORMANCE

ACCELERATION:

Speed range, Gear Ratios and Time in sec.

M.P.H.	4.6 to 1	6.9 to 1	10.5 to 1	19.2 to 1
10—30 ..	—	9.3	6.9	—
20—40 ..	16.3	10.5	—	—
30—50 ..	22.0	15.9	—	—

From rest through gears to:

M.P.H.	sec.
30	7.3
40	13.0
50	22.4

Standing quarter mile 24.5 sec.

MAXIMUM SPEEDS ON GEARS:

Gear		M.P.H.	K.P.H.
Top ..	(mean)	61.6	99.1
	(best)	62.8	100.0
3rd ..	.. ..	52.0	83.7
2nd ..	.. ..	36.0	57.9
1st ..	.. ..	20.0	32.2

TRACTIVE EFFORT:

	Pull (lb per ton)	Equivalent Gradient
Top	150	1 in 14.9
Third	240	1 in 9.3
Second	350	1 in 6.3

BRAKES: (at 30 m.p.h. in neutral)

Pedal load in lb.	Retardation	Equivalent stopping distance in ft.
25	0.15g	201
50	0.46g	66
75	0.85g	35.5
90	0.95g	31.8

FUEL CONSUMPTION:

M.P.G. at steady speeds:

M.P.H.	Direct Top
30	85.1
40	62.5
50	55.6

Overall fuel consumption for 1,135 miles, 45.6 m.p.g. (6.2 litres per 100 km.).

Approximate normal range 44-55 m.p.g. (6.4-5.1 litres per 100 km.).

Fuel: Commercial grades.

TEST CONDITIONS: Weather: Dry, negligible wind.

Air temperature: 56 deg. F.

Acceleration figures are the means of several runs in opposite directions.

Tractive effort obtained by Tapley meter.

Model described in *The Autocar* of 6 September 1957.

SPEEDOMETER CORRECTION: M.P.H.:

Car speedometer:	10	20	30	40	50	60
True Speed:	8.5	18	27	36	45	54.5

DATA

PRICE (basic), with saloon body **£449.**
British purchase tax, **£227.**
Total (in Great Britain), **£676.**

ENGINE: Capacity, 585 c.c. (35.7 cu in).
Number of cylinders: 2, horizontally opposed, air-cooled.
Bore and stroke: 74 × 68 mm (2.91 × 2.68in).
Valve gear: o.h., pushrods.
Compression ratio: 6.8 to 1.
B.H.P.: 19.5 (nett) at 4,000 r.p.m. (B.H.P. per ton laden 13.65).
Torque: 28.9 lb ft at 2,500 r.p.m.
M.P.H. per 1,000 r.p.m. in top gear: 12.4.

WEIGHT: (With 3 gals fuel): 11 cwt (1,232 lb).
Weight distribution (per cent): F, 39.8; R, 60.2 (unladen).
Laden as tested: 14 cwt (1,568 lb).
Lb per c.c. (laden): 2.69.

BRAKES: Type, Ate, hydraulic.
Drum dimensions:
F, 7.1 in diameter; 1.25 in wide.
R, 7.1 in diameter; 1.25 in wide.
Lining area: F, 33.5 sq in; R, 33.5 sq in (95.7 sq in per ton laden).

TYRES: 5.20-10 in Metzeler
Pressures (lb sq in): F, 15; R, 22 (1-2 occupants) F, 17; R, 25 (3-4 occupants).

TANK CAPACITY: 5 Imperial gallons.
Oil sump: 3.5 pints.

STEERING: Turning circle:
Between kerbs, 27ft 7in R, 30ft 4in L.
Between walls, 28ft 2in R, 30ft 11in L.
Turns of steering wheel from lock to lock: 2.5.

DIMENSIONS: Wheelbase, 5ft 7in.
Track: F, 4ft 0in; R, 3ft 9.6in.
Length (overall): 9ft 6in. Width: 4ft 7in.
Height: 4ft 6in. Ground clearance: 6.5 in.

ELECTRICAL SYSTEM: 12-volt; 31 ampère-hour battery.
Head lamps: Double dip; 45-40 watt bulbs.

SUSPENSION: Front independent, leading arms with vertical coil springs.
Rear: independent, trailing arms with coil springs.

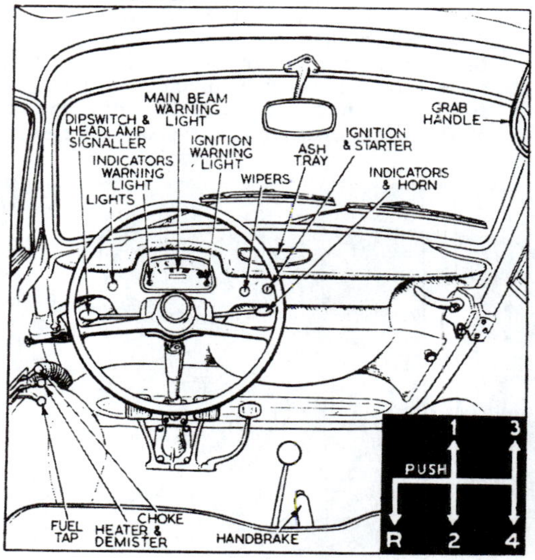

TWO MIDGET cars worthy of attention by those interested in errand-type transportation where long distances are not involved are the BMW Isetta 600 and BMW Isetta 300. Both are extremely economical to operate; both are so compact that they offer tremendous advantages over the ordinary American car and even most other imports when it comes to parking and handling in congested traffic.

Most practical and better of the two is the BMW 600, a comparatively recent newcomer to the import field. Selling below almost all other imports, it has an intriguing initial cost of $1395 at port-of-entry. True, it is tiny in size as compared to say a Plymouth (its wheelbase is 67 inches; Plymouth's 118—overall length is 115; Plymouth's 204!), yet it is entirely capable of carrying four (or five) passengers comfortably. By comfort is meant that there's plenty of headroom, legroom, hip room for four; with three people in the second seat, however, it's a tight squeeze.

Entry and exit to the front seat of the BMW 600 is unconventional—through the front-opening door. The technique of getting in and out is different, but also exceptionally easy once it is learned. The driver steps in standing upright, turns around also while standing upright, then sits down. If the passenger has preceded him, this procedure is more difficult. There's a tendency to grasp the double-jointed steering column and wheel (attached to the door) for support, but this will only swing the door in on him before he is ready for it.

POWERPLANT on BMW Isetta 600 model is rear-mounted, easily accessible. It's an air-cooled, two-cylinder, four-cycle unit that uses a 12-volt starter-generator for starts.

Road Testing The ISETTA 300 AND 600

SMALLER ISETTA 300 MODEL (LEFT) IS ONE-DOOR TWO-SEATER; 600 IS FOUR-PASSENGER (RIGHT).

VERTICAL PLEATED upholstery on test 300 car is custom. Double-jointed steering column and wheel is attached to door; swings with it when opened for entry, exit.

ISETTA 600 has door on right side only for access to rear seat. Long-legged rear seat passengers may have knee-room trouble; otherwise, economy car has good ride comfort.

Once seated, it's a simple matter to close the door by pulling back on the steering wheel, but the reach to the far right to latch the door is awkward. Driving position is comfortable with controls within easy reach; instruments (speedometer and warning lights) can be easily seen. Biggest objection to the controls is the positioning of the tiny clutch and brake pedals, set too close to the steering column. Occasionally the driver's shoe soles would catch on the steering column restricting engagement of the clutch or brake pedal.

Starting the air-cooled two-cylinder, four-cycle engine through a combined 12-volt starter-generator setup is silent and immediate. Shifting the four-speed fully synchromesh gearbox by means of a floor-mounted lever on the right side of the driver is easy. Acceleration, while certainly not like that of bigger-engined cars (the 600 only has 26 hp), is good enough to give the driver confidence in most traffic situations.

The ride in this 1166-pound car is pleasantly comfortable, with less bounce than might be expected in so light a vehicle. This probably can be attributed to the excellent four-wheel independent suspension setup, with coils at each wheel. The front suspension employs a form of rolling arms, a sort of cross between the VW and the '35 Chevrolet. Parts are very light, however, so it does not suffer from the disadvantages of excess weight that the Chevy did.

Steering is positive and quick; there's no feeling of instability in the corners.

LARGER 600 MODEL is comparative newcomer to the import field. Despite its tiny 67-inch wheelbase, car has interior room that is surprisingly adequate for the most part.

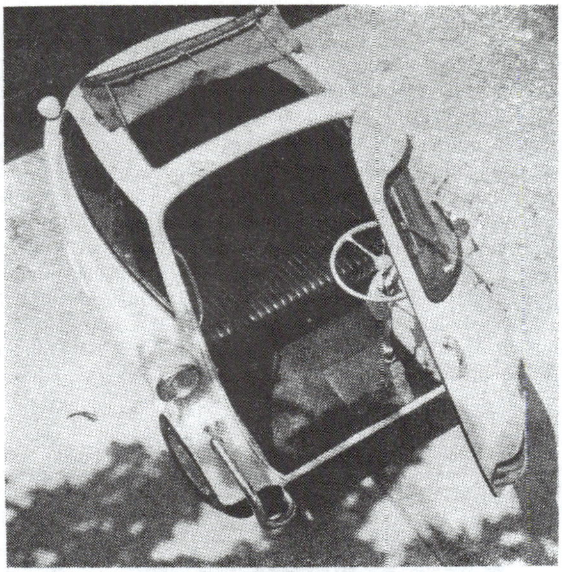

BOTH MINIATURE VEHICLES are available with a canvas fold-back roof; this helps to eliminate feeling of claustrophobia. Test cars were from Santa Monica Imports.

With a turning circle diameter of only 26 feet, the 600 is extremely maneuverable in tight situations. Parking is a downright pleasure.

Legroom in the rear seat is definitely restricted, especially for tall adults. Height of the seat, however, is good, so that the knees of the passengers need not be pressed against their chests.

The smaller BMW Isetta 300 is only a single-front-door, two-seater selling for $1095 at port-of-entry. Weighing 770 pounds without driver, its single-cylinder, air-cooled engine of 13 horsepower, although tiny and noisy, manages to give the driver a feeling of adequate acceleration, possibly because of the vehicle's smallness. Actually, the car will arrive at a speed of 30 mph in about the same time it takes an American high-powered car to get to 60 mph—in a little over 10 seconds. Top speed is only 52 mph as compared to better-than-60 mph for the BMW 600.

Isetta 300

Test Car: 1958 BMW Isetta 300
Body Type: two-passenger, single front-door coupe
Basic Price: $1095 port-of-entry
Engine: fan-cooled, single cylinder, four-stroke
Carburetion: single downdraft
Displacement: 18.1 cubic inches
Bore & Stroke: 2.83 & 2.87
Compression Ratio: 7.0-to-1
Horsepower: 13 @ 5800 rpm
Horsepower per cubic inch: .72
Torque: 14 lbs. ft. @ 4200 rpm
Weight: 770 lbs. without driver
Power-weight Ratio: 59.23 lbs. per horsepower
Transmission: four-speed stick shift
Final Drive: adjustable chain drive in oil bath
Dimensions: wheelbase 58 inches, overall length 93.7, width 55.5, height 52.7, tread 47.2 front, 20.4 rear.
Suspension: independent front wheel suspension, swinging arms, coil springs and hydraulic shock absorbers. Quarter elliptic leaf springs at rear with telescopic hydraulic shock absorbers
Tires: 4.80 x 10
Gas Mileage: 57.2 mpg average
Speedometer Error: indicated 30 and 45 mph are actual 30 and 44 mph, respectively
Acceleration: 0-30 mph in 10.4 seconds, 0-45 mph in 24.4 seconds, 0-52 mph top speed in 38 seconds

Isetta 600

Test Car: 1958 BMW Isetta 600
Body Type: four/five-passenger, two-door sedan
Basic Price: $1395 port-of-entry
Engine: air-cooled, two-cylinder, horizontally opposed four-cycle
Carburetion: downdraft with accelerator pump
Displacement: 35.68 cubic inches
Bore & Stroke: 2.91 & 2.68
Compression Ratio: 6.8-to-1
Horsepower: 26 @ 4000 rpm
Torque: 29 lbs. ft. @ 2500 rpm
Horsepower per cubic inch: 72.9
Weight: 1166 lbs. without driver
Power-Weight Ratio: 45.84
Transmission: four-speed with all forward gears full synchromesh
Rear Axle Ratio: 5.43
Dimensions: wheelbase 66.93 inches, overall length 115, width 55⅛, height 54⅛, tread 48.6 front, 45⅔ rear
Suspension: independent swing arm front and rear, coil springs double-acting shock absorbers
Tires: 5.20 x 10
Gas Mileage: 38.1 mpg average
Speedometer Error: indicated 30, 45, and 60 mph are actual 30, 43, and 57 mph respectively
Acceleration: 0-30 mph in 6.9 seconds, 0-45 mph in 15.2 seconds, 0-60 mph in 34.6 seconds

Getting in and out of the BMW Isetta 300 is similar to entering the front seat of the 600 model. Vision also is the same—excellent all around except for rather thick corner posts made necessary by the front-hinged doors. Driving position is quite similar to the BMW 600—comfortable with adequate head and legroom; foot pedals are positioned too close to the steering column and are too small—same as the 600.

The four-speed shift lever, mounted to the left of the driver, is quite difficult to operate not only because of the awkward positioning but also because of trouble in getting the lever into the right gear. The driver was constantly leaving a stoplight in high rather than in low because of this difficulty; this trouble, incidentally, was not experienced at all with the larger model car.

The ride of the 300 cannot be described as comparable to the 600; it simply wasn't as smooth nor was there the feeling of stability that seemed to be inherent in the 600 car.

For sheer economy of operation, however, the BMW 300 had one distinct advantage over the 600—gas consumption tests revealed a 57.2-mpg average for the two-seater as against a 38.1-mpg average for the four-passenger car. Both of these test figures, incidentally, are somewhat below factory claims of 62 mpg for the 300, 58 mpg for the 600.

Either car can be recommended for specialized uses where minimal transportation and economy are of prime importance. Both are well built in traditional German precise assembly work. They do meet a growing need for a vehicle that is completely compact, efficient, and cheap to operate. •

SCI ROAD TEST BMW 600

No the 600 is not an orthodox approach economy transportation. One thing that the front opening door forced on the designers—high-mounted head lights.

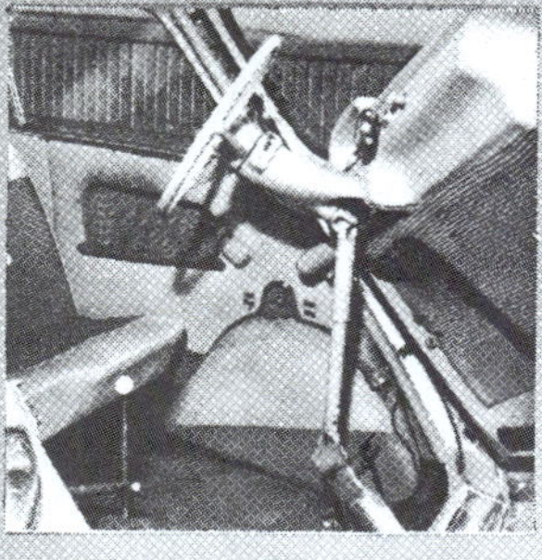

BMW has a two-door body. Of course, one is hinged on the nose. If that isn't enough for the people who like the unusual the complete steering wheel and column swings up.

No matter how you look at it (above) BMW 600 is functional rather than "pretty". Side door (right) is for the use of rear seat passengers only.

► However you expect to view BMW's 600 you must resolve for yourself and your purposes the suitability of the nose hatch. I don't intend to dwell on this aspect of the car, but it is certainly the BMW's most controversial feature and, indeed, the key to the car's whole layout. It should be obvious that the choice of the forward-opening front door for this vehicle (introduced in August, 1957) directly followed the availability of the door parts the Munich firm was already stamping out for its Isetta two-seater. BMW wanted to get into the burgeoning market for 600 cc cars at minimum cost, so the front door, front suspension and general constructional layout of the originally-Italian Isetta were transposed to a four-seater and powered by a detuned version of one of BMW's famed opposed-twin cycle engines.

You'll recall that the door arrangement on the Isetta was eminently justified by the very brief length of the car, which allowed it to be parked with nose to the curb so that driver and passenger could step right out onto the sidewalk. The much longer 600 is denied this privilege. Exit and entry must be made to and from street level, requiring no short step either way. There's also the slim but extant possibility that a car ahead may park close enough to make it difficult to open the door at all. Though its function is now less clear, then, the 600's front door is certainly well executed. Strong spring counterbalancing eases opening while a positive locking motion is necessary to clamp the husky latch shut. Stepping through the opening is not in itself difficult—it is aided by a convenient hand hold—but a fog light, as fitted to the test car, can be a real obstacle. It's also uncomfortable for the passenger to remain seated while the driver leaves or enters.

Several times during the test, onlookers remarked that they "just wouldn't trust a steering column that bent like that". The column in fact is deeply splined, heavily jointed and free from play or other idiosyncracies. Those concerned about the crash hazards of the extreme forward seating will

Dashboard contains only one instrument—a speedometer. Other information is transmitted to the driver by winking lights. Jointed steering column is deeply splined, free from play

be heartened by the very heavy tubular framing around the door opening (not to mention the sturdy front bumper) and by the subtle concealment of the spare wheel and tire within the door.

Though modestly cushioned, the seats are well shaped and angled, the back of the front seat having a distinct twin-bucket contour. An adjustment by wing-nuts over a five-inch range gives ample foot room for any driver, while head-room is better than in most full-sized sedans. The same goes for the back seat, which is some four inches wider than the front one, and which boasts leg room better than that in certain German cars of twice the displacement. There's suit-case-sized luggage space behind the seat which can be greatly extended by folding the seat back forward to form a load-carrying platform finished in rugged cloth. The single right rear door opens wide and allows comfortable access; it's lockable from the inside only. Its window is the only major one that doesn't open, while the driver's pane slides open considerably farther than the right front and left rear.

Expectations of a bus-like driving position are put to rest by the well-angled position of the rather small wheel. The sole instrument is a speedometer which also houses lights for "generator charge" and the directional signals. The latter are activated by a lever at the right of the column, an upward tug on the same stick serving to sound the horn. A similar prong on the left dims and flashes the headlights. Handy to the driver's left hand is a cluster of round-knobbed levers including the choke, the off-on-reserve fuel tap (turn off for extended parking to prevent drip from the carb), and the heater control. Of heat there is aplenty, ducted direct to the feet of the front and rear occupants and to the left side of the windshield for fast and thorough defrosting. Vision all around is very good, with the exceptions of the thick front pillars and the rear view mirror, which can block the forward view of a tall driver. Altogether the driver feels very much at home and in control of his vehicle inside the 600.

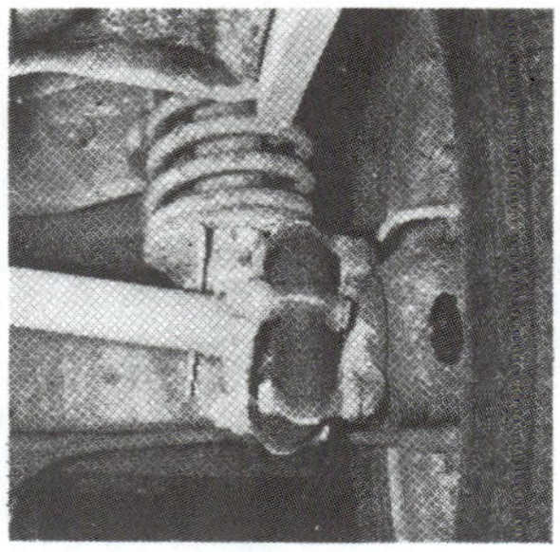

Minor engine controls (far left) are conveniently placed near the driver's left hand. Rear suspension utilizes rubber universal joints just inboard of the wheel hubs.

BMW two-lunger is restricted to 19.5 bhp to come below line of German tax on horsepower. Competition 600s (!) run two carbs and many mods to belt out a solid 35 bhp

Air-cooled twin (above) is pretty well hidden by light sheet metal shrouds. Louvers (left) in back deck admit engine air.

BMW 600

Price at East Coast POE ... $1398
Price at West Coast POE ... $1495

U.S. Importer: Fadex Commercial Corp.
487 Park Ave., New York 22, N. Y.

PERFORMANCE

TOP SPEED:
65 mph

ACCELERATION:

From zero to	seconds
30 mph	6.7
40 mph	11.4
50 mph	19.3
Standing ¼ mile	24.7
Speed at end of quarter	54 mph

SPEED RANGES IN GEARS:

I	0-18
II	7-32
III	11-50
IV	17-top

SPEEDOMETER CORRECTION:

Indicated Speed	Timed Speed
30	28
40	38
50	47
60	57

FUEL CONSUMPTION:
Hard driving ... 46½ mpg

SPECIFICATIONS

POWER UNIT:

Type	Air-cooled opposed twin
Bore & Stroke	2.91 x 2.68 in (74 x 68 mm)
Stroke/Bore Ratio	0.92/1
Displacement	35.7 cu in (582 cc)
Compression Ratio	6.8/1
Carburetion by	One sidedraft Zenith KL/P3
Max. Power	19.5(din) bhp @ 4500 rpm (23 SAE)
Max. Torque	28.5 lbs-ft @ 3000 rpm

DRIVE TRAIN:

Transmission ratios		overall ratio
I	3.45	(18.75)
II	1.94	(10.53)
III	1.27	(6.89)
IV	0.846	(4.70)

Final drive ratio 5.43
Axle torque taken by engine mounts

CHASSIS:

Ladder-type tubular frame

Wheelbase	67 in
Tread, front and rear	48, 45¾ in
Front Suspension	Dubonnet leading arms, coil springs
Rear Suspension	Single trailing arms, coil springs
Shock absorbers	telescopic
Steering type	worm gear
Steering wheel turns L to L	2⅓
Turning diameter, curb to curb	28 ft
Brakes	7.1 in drums
Brake lining area	67 sq in
Tire size	5.20 x 10

GENERAL:

Length	114 in
Width	55 in
Height	54 in
Curb Weight	1230 lbs
Weight, as tested	1320 lbs
Weight distribution, F/R as tested	43/57 (39/61 empty)
Fuel capacity	5⅜ U.S. Gallons

RATING FACTORS:

Specific Power Output	0.55 din bhp/cu in (0.64 SAE)
Power to Weight Ratio, as tested	68 lbs/hp (din) (57 SAE)
Piston speed @ 60 mph	2120 ft/min
Braking Area, as tested	101 sq in/ton
Speed @ 1000 rpm in top gear	12.6 mph

To conserve space in the original Isetta, the Dubonnet "knee-action" suspension was adopted, in which the vertical coil spring and parallel leading arms are carried outboard of the steering king-pin. It still serves in the 600 to keep the wheel housings small and thus improve foot room, but the allowable wheel travel is limited and frequently baffled by bigger-than-usual bumps. As would also be expected, the small wheels (space-saving again) "find" more bumps to be baffled by, but their light unsprung weight allows the shocks to keep them under control. The ride is probably best in the back seat, just between the wheels, while the front seat occupants do sit almost over the front wheels.

When this little BMW gets well under way its all-independent suspension proves very roadable indeed, lending surprising stability and controllability to a car which is unorthodox even by German standards. A small anomaly is a fundamental understeer in spite of 57 percent of the weight on the back wheels with the driver aboard, explained at least partially by the ground-level roll center and wider tread at the front. It lends the car a stubborn directionality which allows corners to be entered with great abandon. If the abandon is excessive and can't be compensated by the excellent steering lock, the tail can be brought around at the limit. The forward driving position is no handicap at all to good handling, but there is a certain top-heavy feel that starts to become important when the outer front wheel (in a corner) nears the end of its modest travel.

With the Dubonnet layout the steering is fully suspended, so the simple linkage is completely isolated from road shock and can be geometrically perfect. The very smooth feel that results escapes being "dead" in this case, thanks in part to the strong caster action that's been supplied. Only slightly less impressive are the brakes, which are so progressive in action as to feel underpowered at first acquaintance. They're up to the car's performance but not much more. Hard applications at low speeds, as in traffic, can cause alarming nose-dipping, though always in a straight line.

It's not strictly true to say that the 600's powerplant is stolen directly from the BMW cycle department. The famed R68/69 series gets a displacement of 592 cc with a bore and stroke of 73 x 72 mm, while the 600's dimensions of 74 x 68 mm reveal that the latter uses the crankshaft of the 494 cc R50 with a bigger bore than the R68 to get a capacity of 582 cc. The cycles deliver 28 and 35 horses in touring and sports tune, while the 600 is held to 19.5 bhp to suit the German horsepower tax situation. For rallies and suchlike BMW goes to twin carbs and many (permissible) mods to get about 35 bhp from the 600 engine.

Starting is a silent cinch with the Bosch Dynastart unit, though the choke must be deactivated *instanter* once the twin pops into life. Aided by flexible mounting and remote location, the smoothness of the BMW's engine is really impressive. Its idle is tangible but not obtrusive, while it has but one important vibration point very low down in its speed range. On the road it makes itself heard as a pleasing Porsche-like rumble from the rear, promising much more power than is there but not offending in doing so. As is usual in such detuned rigs the 600's cruising speed is distinguishable from its maximum only by a heavier throttle spring and a couple of miles per hour, 60 being maintainable day in and day out plus a bonus on any downhill stretch.

Of course acceleration is far from breathtaking but the gearing is so good and the engine so willing that there's a combination for every circumstance. Synchromesh gives quick access to each of the four well-spaced cogs, the lever being conveniently close to the driver's right knee—so close in fact that the knob conflicts with the right shin in reverse and second slots. Most driver will find that third gear will be as high as they care to go in downtown traffic, since fourth gets rough below about 20 mph. The clutch is smooth and decisive on starts but is less equal to the repeated speed shifts I imparted during performance testing or would impart in a race.

The BMW folk proudly say that the 600 is "bigger inside than outside". It's certainly laid out to make use of every square foot of ground it covers, and the resulting shape is, if anything, more efficient aerodynamically than the conventional form. The only items within the car that give an impression of smallness are the foot pedals, which take some getting used to even for confirmed small-car drivers. Capable, solid and sincere, BMW's 600 makes not one claim that it can't back up to the hilt. *—KL*

JOIN THE MICROCAR & MINICAR CLUB!

HOW TO JOIN:

The Microcar & Minicar Club, Inc., is a not-for-profit social club organized to preserve information about microcars and minicars, and to disseminate this information to ensure ongoing enjoyment of small and unusual motor vehicles. Membership in the Microcar & Minicar Club is $20 for one year. This includes four issues of the club magazine, Minutia. Overseas rate is $30. It is not necessary to own a vehicle to be a member. To join, fill out this form and send it with your check to: The Microcar & Minicar Club c/o Richard Bachmann, 2 Second St., North Arlington, NJ 07031 USA. Make checks payable to: The Microcar & Minicar Club.

The Motor Road Test No. 37/58

Make: B.M.W. **Type:** 600

Makers: Bayerische Motoren Werke, 76, Lerchenauerstrasse, Munich 13, Germany.

Concessionnaires: Isetta of Great Britain Ltd., Locomotive Works, New England Road, Brighton, Sussex.

Test Data

World copyright reserved; no unauthorized reproduction in whole or in part.

CONDITIONS: *Weather: Cool and foggy with little wind. (Temperature 37°—42°F., Barometer 30.3 in. Hg.) Surface: Damp tarred macadam. Fuel: Standard-grade pump petrol (approximately 82 Research Method Octane Rating).*

INSTRUMENTS. (Kilometre calibrations)

Speedometer at 30 m.p.h.	10% fast
Speedometer at 60 m.p.h.	10% fast
Distance recorder	3% fast

WEIGHT

Kerb weight (unladen, but with oil, coolant and fuel for approx. 50 miles)	10¾ cwt.
Front/rear distribution of kerb weight	39½/60½
Weight laden as tested	14¼ cwt.

MAXIMUM SPEEDS

Flying Quarter Mile

Mean of four opposite runs	59.2 m.p.h.
Best one-way time equals	60.4 m.p.h.

"Maximile" Speed. (Timed quarter mile after one mile accelerating from rest.)

Mean of four opposite runs	58.7 m.p.h.
Best one-way time equals	60.8 m.p.h.

Speed in Gears

Max. speed in 3rd gear	51 m.p.h.
Max. speed in 2nd gear	35 m.p.h.
Max. speed in 1st gear	20 m.p.h.

FUEL CONSUMPTION

68.0 m.p.g. at constant 30 m.p.h. on level.
59.5 m.p.g. at constant 40 m.p.h. on level.
53.5 m.p.g. at constant 50 m.p.h. on level.

Overall Fuel Consumption for 786.8 miles, 17.4 gallons, equals 45.2 m.p.g. (6.25 litres/100 km.)

Touring Fuel Consumption (m.p.g. at steady speed midway between 30 m.p.h. and maximum, less 5% allowance for acceleration) .. 53.7 m.p.g.

Fuel tank capacity (maker's figure) .. 5 gallons

STEERING

Turning circle between kerbs:

Left	27 feet
Right	31¼ feet
Turns of steering wheel from lock to lock	2¼

BRAKES from 30 m.p.h.

1.00g retardation (equivalent to 30¼ ft. stopping distance) with 75 lb. pedal pressure.
0.80g retardation (equivalent to 37¾ ft. stopping distance) with 50 lb. pedal pressure.
0.32g retardation (equivalent to 94 ft. stopping distance) with 25 lb. pedal pressure.

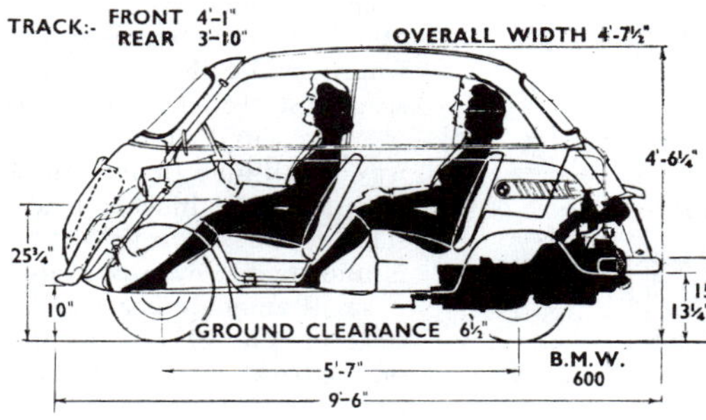

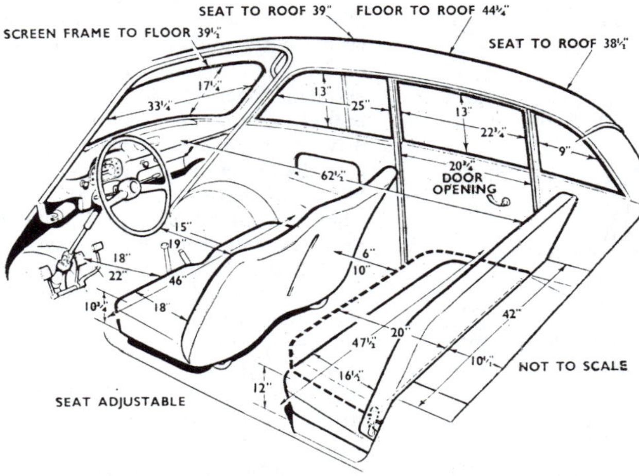

ACCELERATION TIMES from standstill

0-30 m.p.h.	7.6 sec.
0-40 m.p.h.	14.2 sec.
0-50 m.p.h.	24.8 sec.
Standing quarter mile	25.6 sec.

ACCELERATION TIMES on Upper Ratios

	Top gear	3rd gear
10-30 m.p.h.	16.7 sec.	8.5 sec.
20-40 m.p.h.	16.9 sec.	10.7 sec.
30-50 m.p.h.	22.5 sec.	16.2 sec.

HILL CLIMBING at sustained steady speeds

Max. gradient on top gear	1 in 12.5 (Tapley 180 lb./ton)
Max. gradient on 3rd gear	1 in 6.9 (Tapley 320 lb./ton)
Max. gradient on 2nd gear	1 in 5.1 (Tapley 430 lb./ton)

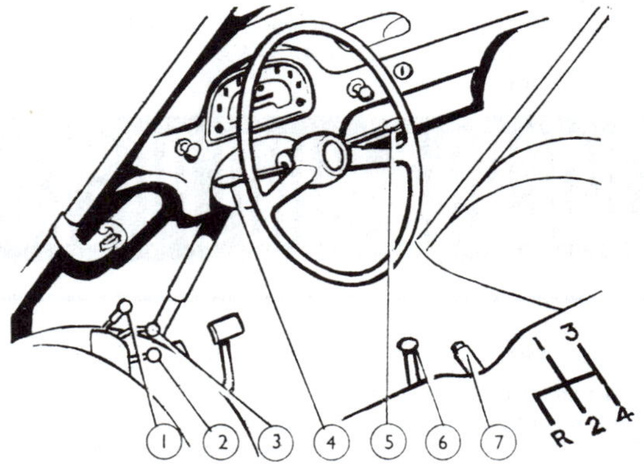

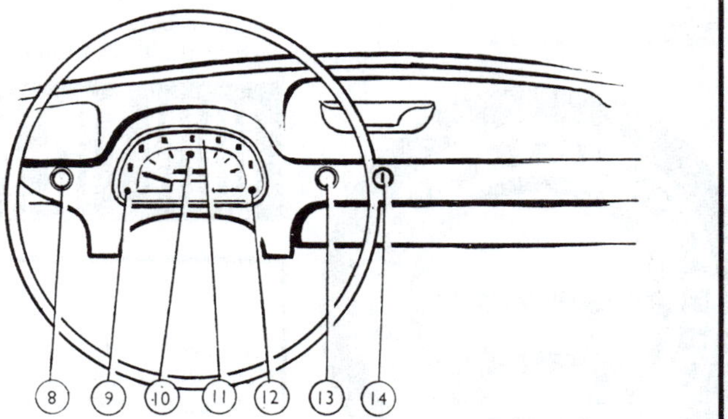

1, Choke control. 2, Heater control. 3, Petrol and reserve tap. 4, Headlamp flasher and dipper. 5, Horn and direction indicator switch. 6, Gear lever. 7, Handbrake. 8, Lights switch. 9, Direction indicator warning light. 10, Headlamp main beam indicator light. 11, Speedometer and distance recorder. 12, Dynamo charge warning light. 13, Windscreen wiper switch. 14, Ignition and starter switch.

The B.M.W. 600

SMALL by comparison with London taxis, and even with other small private cars, the B.M.W. 600 nevertheless is by no means a "miniature" but offers full four-seat motoring by use of compactly novel layout. On the car tested, the front-opening door could have been improved by stronger counter-balancing to reduce the effort of leaving the car.

A Lightweight German Mile-a-Minute Four-seater of Exceptional Compactness and Operating Economy

THE most obviously distinctive thing about the B.M.W. 600 which we have recently been testing on British roads is that it is 22 inches shorter overall than any 4-wheeled 4-seater car which is made in Britain. What is not always appreciated without practical experience is the fact that, despite its small size, this is in every sense a "real" car and should not be considered as a miniature. Only available in Britain as an import from Germany, the B.M.W. 600 is expensive here at present because import duty is added to the cost of a car from a Munich factory more famous for engineering quality than for rock-bottom prices. As an example of successful unconventionality, however, its interest goes far beyond the limits set by its present price to British sales.

The layout of this unique little car is well seen in our usual diagram on the opposite page. Two people occupy a seat which is only slightly behind the front wheels, reaching this seat through a forward door of the kind which "bubble cars" inherited from the hansom cab of an earlier century. Two more people can enter a conventional rear seat through a conventional side door, and behind them there is a useful luggage well, the capacity of which can be multiplied by folding down the rear seat backrest in "estate car" fashion. Behind the rear wheels, a twin-cylinder horizontally-opposed engine of fan-cooled o.h.v. design is located, its not-too-small 585 c.c. size allowing good torque to be combined with the 19½ b.h.p. maximum output which ensures a top speed of just about a mile a minute. Built in simple but durable-seeming fashion, this miniature vehicle goes on to the scales at a ready-to-drive weight of only 10¾ cwt.

In terms of figures, the most impressive thing about this little B.M.W. is its economy of running costs. Engine oil consumption during our test was virtually nil, the cheapest grades of fuel sufficed to keep the engine free from pinking, yet the overall petrol consumption of 45.2 m.p.g. is notably superior to any four-seat saloon which we have sampled which offers at all comparable performance. As usual, we drove this little car hard for a major part of the time that it was in our hands, making free use of the four-speed gearbox to obtain rapid acceleration, although a strong "second pressure" in the accelerator linkage does encourage a driver to ease back from full throttle to a setting which allows a 50-55 m.p.h. cruising speed to be held economically on open roads.

The well-balanced design of this small car has been happily successful in allowing good cruising fuel economy figures (such as 59½ m.p.g. at a steady 40 m.p.h.) to be secured without resort to an exaggeratedly high top gear. Extreme top gear liveliness is not to be expected, but there is pleasantly immediate response to the accelerator pedal in top gear over the vital speed range from 15 m.p.h. to 50 m.p.h., especially when the driver is alone but also when a heavy passenger load is being carried. A four-speed gearbox (with synchromesh on all its forward ratios) permits the keen driver to get along at much more than the average speed of main road traffic, 3rd gear being so quiet at a steady 30 m.p.h. that one can easily forget to engage top gear around town, and the red "3" mark at about 40 m.p.h. on the speedometer seeming to err on the side of caution since 50 m.p.h. can be exceeded in this ratio. The central gear lever ought to be convenient, but in fact proves liable to get hidden amongst the coat-tails of an overcoated driver or a passenger in the comfortably-shaped but fairly narrow front seat.

It would be untrue to claim that this air-cooled twin-cylinder four-stroke engine entirely equals the refinement of the best

In Brief

Price in Britain: £449 15s. 4d., plus purchase tax £226 4s. 8d. equals £676.

Capacity	585 c.c.
Unladen kerb weight ...	10¾ cwt.
Acceleration:	
20-40 m.p.h. in top gear ...	16.9 sec.
0-50 m.p.h. through gears	24.8 sec.
Maximum direct top gear gradient	1 in 12.5
Maximum speed	59.2 m.p.h.
"Maximile" speed	58.7 m.p.h.
Touring fuel consumption ...	53.7 m.p.g.

Gearing: 12.15 m.p.h. in top gear at 1,000 r.p.m.; 27.3 m.p.h. at 1,000 ft./min. piston speed.

THE FRONT seats are reached through the opening front panel, to which the double-jointed steering column is attached; gear lever and handbrake are close together in the centre of the floor. A single offside door gives access to the rear seats, the backrest of which can be folded flat to form a luggage platform ahead of the normal baggage well, as seen above.

The B.M.W. 600

EXTREMELY neat detail design is evident, the decorative slats on the rear quarter panels being in fact air intake louvres for the rear engine which is reached through the louvred rear door. Although the horizontally-opposed twin-cylinder engine is fully enclosed in the body there is enough space around it to avoid difficulties in routine maintenance.

water-cooled four-cylinder units, but at most times the disadvantage which accompanies simplicity and freedom from freezing worries is very slight. The tick-over of this engine is smooth, as is its running at speeds above 15 m.p.h. in top gear, and the amount of noise inside the car is very modest during normally economical driving; although at high engine r.p.m. and with "second pressure" applied to the accelerator the power unit becomes much more audible, the same is true of most orthodox small cars. It is only below 15 m.p.h. that slight jerkiness on the over-run asks the lazy driver to change down out of top gear. With the choke correctly used, the engine was instant in starting from cold, in response to a silent dynamo-starter unit controlled from the ignition key, but did not warm up and settle down to operate normally and idle reliably without aid from the choke quite as soon as do some other air-cooled units.

An unfortunate effect which this car shares with some other cars incorporating rear-mounted engines, especially when they are air cooled, is a rather cold interior during winter weather. Provision is made for bringing some of the fan-circulated engine cooling air into the body interior, but when the outside air temperature drops towards freezing point this provides screen de-misting and a very slight degree of interior warmth for one side of the car only. For more temperate weather, three sliding windows permit ample ventilation to be secured without undue draughts, and apart from too-thick windscreen pillars there is a fine all-round view out of the car so long as the windows do not mist up.

There is much to be said in favour of the single forward-opening door which

THE INSTRUMENT panel is on the front door, which also accommodates the spare wheel. Choke, heater and fuel cock controls are on the left-hand body side, headlamp dipswitch lever on the left of the steering column, horn and direction indicator lever on the right.

accompanies an over-the-axle driving position on this car, even when applied to a vehicle long enough to demand normal rather than nose-to-the-kerb parking. It is an especial convenience to be able to pull up beside the kerb and immediately open the door to step out, without any need to await a gap in the stream of passing traffic. On our test model it was unfortunate that spring counter-balancing of the forward door (within the thickness of which the spare wheel is concealed from view) was just not strong enough to hold the door securely open at all times, and entry to or exit from a fairly low seat was further impeded by the fact that use of a more orthodox facia panel than is fitted to the smaller Isettas did not seem to let the steering wheel swing quite so completely out of the way; these however are only alterable details of a fundamentally satisfactory space-saving layout.

Only a single door is provided for access to the rear seat, on the right-hand side of this left-hand-drive German model. The rear seat does not provide much knee-room for big men facing squarely forwards, but headroom is ample and there is so much breadth of floor entirely unobstructed by any transmission hump that by sitting slightly sideways two adults can make themselves reasonably comfortable—one adult can be very comfortable indeed in the back seat, behind which the luggage well is quite usefully roomy. The all-independent coil spring suspension layout used for this model does not at any time give outstanding comfort, the moderately flexible springs not infrequently contacting their flexible buffers when a full load is being carried briskly along a rough road, but whilst no high praise is earned there is equally never any cause to describe the riding of this extra-lightweight vehicle as uncomfortable. We were interested to note that, despite having rather small wheels, it could drive steadily through surprisingly deep mud without loss of traction, thanks to I.R.S. and a rearward centre of gravity.

In some respects the handling qualities of the B.M.W. 600 disappointed us, even after a worn-out rubber bush at one end of the steering drag link had been renewed (the test model had about 17,000 miles recorded on its speedometer) to eliminate lost motion. It soon became natural to take advantage of the compactness and lightness of the vehicle by weaving through traffic in very brisk fashion, but whilst road holding was good the steering remained slightly less light and less quick in response than we would have expected. The hydraulic four-wheel brakes proved quickly responsive and the pull-up hand-brake was above criticism, but hard braking tended to lock one rear wheel and to bottom the front suspension audibly.

Simply furnished in plastics, haircord carpet and rubber matting, the body interior was nevertheless equipped with some useful refinements of detail. There is a good interior lamp, and whilst no contents gauge is provided there is a reserve tap for the 5-gallon petrol tank. Behind the steering wheel there are two finger-tip control levers, that on the right operating the self-cancelling turn indicators and also (when squeezed towards the steering wheel) the horn, that on the left dipping the headlights and also (when squeezed) flashing the headlamp main beam regardless of whether or not any lights are switched on. Faith in the power unit's reliability is reflected by the total absence of any gauge or warning light to indicate oil circulation irregularities.

In recent years, orthodox small cars have gradually been growing up and a new breed of miniature cars has come into being, but in this country a wide gulf has been left between the small cars with over-900 c.c. engines and the miniature cars with engines of under 300 c.c. In Britain, perhaps, the greatest importance of the B.M.W. 600 is, as a reminder to both manufacturers and car buyers that, somewhere midway between these sizes, a compromise can be struck which combines most of the carrying capacity and effortless performance of one with most of the compactness and economy of operation of the other. It is sincerely to be hoped that before long either this or some equivalent car will be built in Britain so that it can be marketed here at a truly competitive price.

The World Copyright of this article and illustrations is strictly reserved. © Temple Press Limited, 1958.

Specification

Engine

Cylinders	2 (horizontally opposed, air cooled)
Bore	74 mm.
Stroke	68 mm.
Cubic capacity	585 c.c.
Piston area	13.3 sq. in.
Valves	Pushrod o.h.v.
Compression ratio	6.8/1
Carburetter	Zenith 28KLP1 horizontal
Fuel pump	None (gravity feed)
Ignition timing control	Fixed
Oil filter	Full-flow
Max. power (net)	19.5 b.h.p.
at	4,000 r.p.m.
Piston speed at max. b.h.p.	1,780 ft./min.

Transmission

Clutch	Single d.p.
Top gear (s/m)	4.6
3rd gear (s/m)	6.9
2nd gear (s/m)	10.5
1st gear (s/m)	19.2
Reverse	18.75
Propeller shaft	Nil (rear engine)
Final drive	7/38 spiral bevel
Top gear m.p.h. at 1,000 r.p.m.	12.15
Top gear m.p.h. at 1,000 ft./min. piston speed	27.3

Chassis

Brakes	Hydraulic
Brake drum internal diameter	7.1 in.
Friction lining area	67 sq. in.
Suspension:	
Front	Dubonnet-type i.f.s. by coil springs
Rear	Independent by trailing links and coil springs
Shock absorbers	Telescopic
Steering gear	Worm and nut
Tyres	5.20—10

Coachwork and Equipment

Starting handle	None
Battery mounting	Under rear seat
Jack	Screw pillar type with ratchet handle
Jacking points	External sockets under sides of body
Standard tool kit	None
Exterior lights: 2 dipping headlamps with pilot bulbs, 2 tail lamps, number plate lamp.	
Number of electrical fuses	6
Direction indicators: Self-cancelling flashers, white front, amber rear, combined with stop lamps.	
Windscreen wipers	2-blade electrical, self parking
Windscreen washers	None
Sun vizors	None
Instruments	Speedometer with non-decimal non-trip distance recorder
Warning lights: Dynamo charge, direction indicators, headlamp main beam.	
Locks:	
With ignition key	Ignition/starter switch and front door
With other keys	None
Glove lockers/map pockets	2 deep wells alongside front seat
Parcel shelves	None (luggage well behind rear seat)
Ashtrays	One on facia panel
Cigar lighters	None
Interior lights	One in roof
Interior heater: Warm air from engine ducted to windscreen interior and front and rear compartments.	
Car radio	None
Upholstery material	Plastic
Floor covering	Rubber mats
Exterior colours standardized: Six single-colour; five two-colour.	
Alternative body styles	None

Maintenance

Sump	3½ pints, S.A.E. 10W/30 (summer and winter)
Gearbox and final drive	2¼ pints, S.A.E. 10W/30 oil
Steering gear lubricant	Grease
Cooling system capacity	Nil (air cooled)
Chassis lubrication: By grease gun every 1,200 miles to 10 points, and check level of 10W/30 oil in i.f.s. reservoirs.	
Ignition timing	Marked on cooling fan, adjacent to contact breaker
Contact-breaker gap	0.016 in.
Sparking plug type	Bosch W225T2 (long reach)
Sparking plug gap	0.027-0.031 in.
Valve timing (with clearance increased to 2 mm.): Inlet opens 4° after t.d.c. and closes 36° after b.d.c.; exhaust opens 36° before b.d.c. and closes 4° before t.d.c.	
Tappet clearances (cold):	
Inlet	0.006 in.
Exhaust	0.006-0.008 in.
Front wheel toe-in	0.080 to 0.115 in. at rims
Camber angle	1½°
Castor angle	16°
Steering swivel pin inclination	5°
Tyre pressures:	
Front	15-17 lb.
Rear	22-25 lb. according to load
Brake fluid	A.T.E.
Battery	12 volt, 31 amp./hr

A CAR ROAD TEST

The BMW 600

PURPOSEFUL originality is the keynote of the BMW 600. Highly unconventional, it is also efficient and entirely practical. Essentially economical, it is able to achieve better than 50 m.p.g. when carrying 4 people at a steady 50 m.p.h.

In designing this car, the manufacturer exploited two major assets : (a) the experience gained with the single-cylinder Isetta 300 model, and (b) the existence within the organisation of a very fine twin-cylinder motor cycle engine, in full production.

Many two-wheel enthusiasts maintain that the twin-cylinder (horizontally-opposed) BMW engine is the finest of all power units for a large motor cycle to have. In motor cycle form (with twin carburetters) this engine develops about 30 b.h.p. Duly "tamed", or detuned, for use in the 4-seater BMW 600 (with a single carburetter, reduced compression ratio, revised valve timing, etc.) it yields 19.5 net (23 gross) b.h.p. German insurance rates favour cars with engines developing less than 20 (net) b.h.p., and the detuning process has been carried out to an extent which has resulted in an air-cooled twin-cylinder engine of notable smoothness that is capable of producing good torque at quite low r.p.m.—in fact, a unit very well suited to providing modern ultra-small-car motoring.

Transmission is essentially car-type. The drive is taken via a single-dry-plate clutch to a gearbox that provides 4 forward speeds, all equipped with synchromesh, and thence to a rear axle that incorporates differential gearing and provides a reduction of 5.43 to 1 (spiral bevel pinion and ring gear). The gearing that is provided by the gearbox is shown in the data panel at the end of this report; all ratios are indirect, and their overall values are : 1st 19.22 to 1, 2nd 10.53 to 1, 3rd 6.90 to 1, top 4.59 to 1.

An overall gearing of nearly 4.6 to 1 for "top" may appear to be somewhat high in view of the engine's modest cubic capacity (584.9 c.c.), but the tiny wheels are shod with 5.20-10 tyres. The final results are that a crankshaft speed of 5,000 r.p.m. provides a road speed of 60 m.p.h. and that the BMW 600 proves to be capable of pottering as slowly as 15 m.p.h. in top gear.

There are, however, sound reasons for making good use of the gear lever : the ratios are excellently chosen to match the engine's characteristics; the centrally-positioned lever has a pleasingly short and positive movement; habitually making the engine pull at too low a speed in too high a gear is unwise because neither the output of the generator nor the volume of air that is displaced by the centrifugal engine-cooling fan is particularly generous at low r.p.m. Then, of course,

performance is greatly influenced by the use that is made of the gearbox: 20-40 m.p.h. takes 16 seconds in top, compared with a trifle over 10 seconds in 2nd gear.

The manufacturer recommends the following maximum and cruising speeds: 1st 15 m.p.h., 2nd 28 m.p.h., 3rd 44 m.p.h., top 62.5 m.p.h.

As usual, we timed maximum speed on a level road and took the average of several runs in opposite directions; the result of this was 61.8 m.p.h. It seems clear that the recommended maxima for 2nd and 3rd gears have been conservatively selected to ensure long, trouble-free engine life, for the test car was very willing to run up to 33 m.p.h. in 2nd and 50 m.p.h. in 3rd. Long, trouble-free life should, indeed, be a feature of this engine which has been so thoroughly proved in more highly-stressed form.

Two distinct pressures are encountered in the travel of the accelerator pedal. The first pressure operates during approximately the first $\frac{1}{3}$-travel, thereafter the driver is called upon to exert appreciably more effort. On flat stretches of open road, the early part of the travel provides an easy cruising speed of 50-55 m.p.h. with commendably little noise. The noise level rises when the accelerator is depressed further, and, as they are closer to its source, rear-seat passengers are more aware of it than are the occupants of the front seat.

Obviously, not only the quietest but also the most economical motoring is obtained when the accelerator is not depressed beyond the range of the first-pressure. Driven in this fashion, the BMW 600 can return m.p.g. figures well in excess of 50. Making fuller use of the available performance—and including in our test a number of short journeys, for this type of car is often

The two flanks of the car are completely different. The off-side incorporates a door serving the rear seat; the near-side has no opening. Ventilation is obtained by sliding-windows, that next to the driver slides sufficiently far back to provide an opening through which he can thrust his head when reversing.

The upper parts of the engine (the cylinders are beneath the floor of the engine compartment) remain remarkably clean. Fuel is fed by gravity from a 5-gallon tank mounted immediately above the engine. An on-off-reserve tap, controlled from the driving compartment, is fitted; the Instruction Manual warns "Shut off fuel line control tap immediately the engine is stopped."

The rear seat is 48-ins. across at its widest point, the floor is flat, headroom is ample for people up to about 5 ft. 10 ins. tall, but knee-room is somewhat restricted and, to compensate for this, adults tend to sit at an angle.

used for that sort of motoring—we obtained an average of 48.6 m.p.g.

Experiments revealed that the BMW 600 is capable of achieving 51.5 m.p.g. when carrying four adults (the maximum permissible load is 750 lbs.) at a steady 50 m.p.h. on a level road, and that if the cruising speed is reduced to 40 m.p.h. 61 m.p.g. is obtainable in these conditions.

Riding comfort is inevitably limited by the short wheelbase (5 ft. 7 ins.) and by the very small wheels.

(continued overleaf)

The steering column is telescopic and incorporates two universal joints. The spare wheel is carried in a compartment on the inside of the front door. The front seat, secured by two wing nuts, is adjustable. Headroom above the front seat is sufficient for six-footers.

Badly worn tarred surfaces produce quite a considerable amount of vertical motion and occasional (though not marked) pitching. Very little "float" is induced by gently undulating smooth surfaces. Generally, the faster the car is driven the smoother the ride becomes. In part, no doubt, because the sprung/unsprung weight ratio of a very small vehicle like this is appreciably affected by the load that is carried, riding comfort is at its best when all seats are occupied.

Roadholding, on the other hand, is best when the driver is alone in the car or has just one passenger (on the front seat) with him—and it is then quite remarkable. With very little body-roll, and no noise from the tyres, the BMW 600 can be taken round curves and corners with considerable zest. This is one of the things which contributes towards making this little car "fun to drive"—another is the positioning of the controls and the manner in which they operate.

All three pedals are pivoted at their lower ends: they work in accord with ankle-anatomy. The clutch is smooth, and, as previously mentioned, the gear lever operates with admirable precision. Although the Instruction Book states "Snappy changes are liable to beat the synchromesh", we found that, given a reasonable sense of timing, quite rapid changes could be made. The more expert (or should we say old-fashioned?) driver happily discovers that the spacing of the accelerator and brake pedals is such that double-declutch downward changes are readily effected whilst braking for a corner. An oddity to which a newcomer to the car quite quickly becomes accustomed is the fact that the gear lever is spring-loaded towards the *lower* ratios.

Positioned in a manner that shames many far larger cars, the BMW 600's steering wheel enables a six-footer to drive the car without angling his knees to prevent the upper parts of his legs from encountering the rim of the steering wheel when he operates the pedals.

The hand-brake also merits emulation. It is powerful

and easily applied; its efficiency permits it to be viewed as an emergency brake, rather than merely a device to ensure safe parking on an incline.

The main hydraulic braking system is excellent. Its power is fully up to the car's performance and the distribution of effort between the front and rear wheels is such that, whenever necessary, the brakes can be applied heavily with confidence.

Three small controls, placed close to the driver's right hand, control the choke, the feeding of warm air into the car from the engine compartment, and the supply of fuel to the engine (on-off-reserve tap; there is no fuel gauge).

A lever mounted on the left of the steering column provides high and low beams of the headlights, and (when lifted towards the rim of the steering wheel) enables the headlights to be flashed irrespective of the position of the main lighting switch which is on the facia panel. A similar lever on the right of the steering column operates the self-cancelling turn-indicator flashers and the horn.

Ample power for fast night-driving is provided by the main beam of the headlights. We found the warning light of the turn-indicator flashers rather blinding at night; an owner who often motors at night might choose to fit a less powerful bulb here.

The turning circle is larger than might be expected in view of the vehicle's very modest overall dimensions. The steering ratio gives the impression of being exactly right; it provides for $2\frac{1}{3}$ turns of the steering wheel to move the front wheels from one extreme lock to the other. There is good self-centering action, and at no time is any kick-back felt at the steering wheel.

The windscreen surround is massive but the driver is seated so close to the windscreen that his view of the road is not thereby impeded. The windscreen wiper blades work in parallel and are arranged to give preference to the driver's side.

Starting presents no problem. The choke must frequently be employed, for, like all air-cooled units, the engine rapidly loses its heat to the atmosphere after it has been stopped. Starting is virtually silent, for the

All the controls are conveniently positioned. The speedometer dial incorporates an odometer (distances of less than one mile are not indicated) and warning lights for the turn-indicator flashers, main beams of the headlights, and generator-charging. Engine oil pressure is not indicated.

starter motor is combined with the generator and is therefore in constant-mesh.

Weather-sealing of the front door—and, indeed, of the entire vehicle—is good. The only disadvantage of the front door is revealed when a thoughtless motorist parks so close to the front of a parked BMW that the owner of the latter cannot gain access to his locked vehicle. Care is needed when using the single door that serves the rear seat; planned for countries with keep-right rule of the road, it is on the off-side of the vehicle.

The BMW 600 is an extremely economical, neat and well-planned little car that is sturdily constructed and, in view of its price, generously equipped and nicely finished. Its dimensions are diminutive for a 4-seater, but its original layout is so successful that—but for the obvious absurdity of the assertion—one would be tempted to maintain that "it is bigger inside than out"! ●

SPECIFICATION AND PERFORMANCE

BRIEF SPECIFICATION

Make BMW (Bayerische Motoren Werke).
Model 600
Style of Engine Horizontally-opposed twin-cylinder. Air-cooled (centrifugal blower). Overhead valves (push-rods).
Bore ... 2·913 ins. (74·0 mm.)
Stroke ... 2·677 ins. (68·0 mm.)
Cubic Capacity 35·69 cu. ins. (584·9 c.c.).
Maximum Horse-Power 23 b.h.p. (gross) at 4,500 r.p.m. 19·5 b.h.p. (net) at 4,000 r.p.m. (Compression ratio 6·8 to 1).
Brakes ... ... Hydraulic
Front Suspension Independent. Dubonnet-type (swinging arms; leading-link). Coil springs. Telescopic dampers.
Rear Suspension Independent. Coil springs. Swinging arms; trailing-link. Telescopic dampers.
Transmission System Clutch and four manually-engaged forward gears. Synchromesh on all forward gears.
Gear Ratios ... 1st 3·54 to 1
2nd 1·94 to 1
3rd 1·27 to 1
Top 0·846 to 1
Rev. 3·45 to 1
Final Drive Ratio ... 5·43 to 1
Overall Length ... 9 ft. 6 ins.
Overall Width ... 4 ft. 7 ins.
Overall Height ... 4 ft. 6 ins.
Ground Clearance ... ... $6\frac{1}{2}$ ins.
Turning Circle ... 28 ft. approx.
Dry Weight ... ... 1,210 lbs.
Price ... ... £486 at Coast Ports

PERFORMANCE

Acceleration 0-30 m.p.h. 7·5 secs.
0-40 m.p.h. 13·8 secs.
0-50 m.p.h. 22·5 secs.
In top gear from a steady 20 m.p.h. to 40 m.p.h. ... ... 16·0 secs.
In top gear from a steady 30 m.p.h. to 50 m.p.h. ... ... 22·2 secs.
Maximum Speed ... 61·8 m.p.h.
Reasonable Maximum Speed in 3rd Gear 44 m.p.h. (manufacturer's recommendation).
Fuel Consumption ... 48·6 m.p.g.
Test Conditions Sea level. Little wind. Dry road. 90-octane fuel.

CROSS COUNTRY on one cylinder

Hunter and Wieman explored ghost town off beaten track near Montoya, N. M.

By

James A. Hunter

Photos by the author

One of the smallest of small cars, a tiny Isetta "300" proves its stamina on trip from Omaha to Golden Gate

● "Let people laugh, let the snow fall! We'll make the trip regardless!" These were my words to my friend Gene Wieman concerning our upcoming trip to the West Coast in my little Isetta "300". We had discussed the trip for some time but had never made any real concrete plans. All we had decided was that as soon as Gene got out of the Air Force in November we would leave for the Coast. We had so much faith in the "bug" that we knew all we would have to be able to do was read a map.

I had talked to a few Isetta owners around Omaha and they all agreed with me that the car would make it with little or no trouble. They also agreed with me that, since it was so late in the year, we should travel straight south from Omaha and then turn west somewhere in Texas. The Rocky Mountains and the winter snows would be quite an obstacle for the Isetta in the early winter.

The snow had been falling for two days as our departure date neared. We were due to leave Saturday the 29th of November, and on the 26th the snow was coming down steadily. The 28th dawned clear and cold, but sunny and with signs of melting snow. In the crisp winter morning and with high spirits, I took my Isetta down to Le Bron Motors to make final preparations for our junket across the Western states. I took along a list of the things I wanted done so that nothing would be overlooked.

I had visions of exceptional mileage on the trip, so one of the first things I had done was to get my spark plug changed. I had adjusted the brakes myself, and checked the battery water and brake fluid. All three were in good shape, and only the battery required any attention during the entire trip.

Next, the car was greased—only nine simple grease points. Then the transmission, chain drive, and engine were all drained and refilled with S.A.E. 30 oil. The engine will take either detergent or non-detergent oil. That was all there was to it. My spare was in good shape, and the car had already run faultlessly for 2,000 miles. Any newcomers to the small car field should realize that the Isetta has an air-cooled four-cycle engine. Also, it is a "one-lunger," having but a single cylinder.

After spending a hectic evening packing and taking care of all the small problems that invariably turn up, we arose the morning of the 29th to a sunny but cold day. The skies were clear and we were filled with anticipation and anxious to be on the road. The temperature was 10 degrees above

Continued on Page 126

Homemade top carrier and suit-case were taken off to reduce wind resistance during storm in Oklahoma, where Isetta had to "tack" against wind once.

Coming across only the thin Panhandle of Texas made the one-time largest state seem small compared to the others.

Little "300" waits in front of Le Bron Motors in Omaha for pre-trip service check.

Gene Wieman is about to climb in the "bug" at Arizona line.

Continued from Page 124
zero. With one suitcase tied on the luggage rack in back, and the other on our homemade top carrier, we set off.

We planned to go to Kansas City the first day, but after reaching Kansas in less than three hours, decided to push on and see just what the little car was capable of in one day. There was no snow on the roads in Kansas and little traffic, thus we reached Topeka and the Kansas Turnpike at four in the afternoon. The Turnpike turned out to be a real mile eater. In less than four hours we went across Kansas and into Oklahoma.

As we were leaving the Turnpike and entering Oklahoma, we ran out of gas. There is no gas gauge on the Isetta

"300", so when you run out, you merely switch on your reserve tank. The reserve tank is only a copper tubing in the regular gas tank, and allows you from 20 to 40 more miles of driving, depending on the conditions. We soon found that we were only going to get 20 miles or so out of our reserve gas, and the next town was still 19 miles away.

When we reached the small town of Braham, we discovered all the stations were closed for the night. We called one owner, and he sold us 3.5 gallons of gas. The tank only holds 3.4 gallons!

We spent the night in Blackwell, Okla., and upon summing up the first day, we came to these conclusions. The first thing in the morning we would buy a 2-gallon gas can to carry as a spare. Also we would be able to turn out 400 miles a day, and still have a leisurely hour for lunch. We had driven 10 hours the first day and never became cramped or crowded. The visibility for both driver and rider was very good.

On the morning of the 30th, we left Blackwell at 10 o'clock. Still going straight south, we headed for Oklahoma City. The day was clear and sunny but it soon became apparent that we were bucking tremendous headwinds. The car would not go over 40 miles an hour, and even after we took the top carrier off we only picked up a couple of miles an hour. At Oklahoma City we were forced to turn west to get out of the wind. We took Route 66 to Amarillo. The roads were good and we reached the Texas border about 6 that evening.

A four-lane freeway greeted us on our entrance to Texas. It was a straight run of 4 hours to Amarillo, and we got there at 10 that night. Thoroughly pooped, with some 770 miles under our belts in less than 20 driving hours, we felt we had earned the right to sleep in, the next morning. Up to that time, we had used 22.8 gallons of gas, an average of a little better than 33 mpg. Our load and wide-open driving was telling on our gas mileage.

The morning of December 1st brought us sunshine and clear skies again. We did not know it at the time, but our whole trip was to be blessed with this type of weather. The sun was high in the sky when we left Amarillo. The Isetta dealer there had checked over my "bug" and found that only the chain drive and transmission required oil. The engine oil, of course, was added to about every other stop for gas.

As soon as you get out of Amarillo the vast empty land swallows you up, even more so in an Isetta. At Santa Rosa, New Mexico we turned south again on Route 54. We had reached the New Mexico border so fast after leaving Amarillo that we decided to look over New Mexico closely. The land was all rolling and quite empty south of Santa Rosa. It seemed serene and old, as we passed through the towns one by one. Duran, where the Indians were impressed by our car. Then Carrizozo at sunset, and Oscura, and finally Alamogordo, to remind us we were not in the past. For it was only a few miles from there that the first atomic bomb had been exploded. We spent the night in Alamogordo under the shadow of the new and the old.

The morning of the 2nd, determined to really put some miles on the "bug", we got up at 7:00 a.m. After a hurried breakfast, we were soon passing through White Sands National Monument. Then we began to climb up San Augustine Pass, and we had to hold her in 2nd gear all the way, but the little "300" never did falter. At the top, Las Cruces was visible in a fertile valley some 20 miles away.

Hamburgers at 10:00 a.m. in Las Cruces, and we set our sites for Phoenix. Going northwest, we hit Arizona in the early afternoon. Highway 70 led to Safford, and then to the Apache reservation, where we received and returned various war whoops from the Indians. As the sun fell once again, we were entering Phoenix, and were soon surrounded by palm trees and street lights. That night while totaling up expenses, we found that we had spent $13.59 for gas and oil. We had gone 1,570 miles on 44 gallons of gas, which gave us an average of 36 mpg. Nothing to brag about, but it was good.

The next day was so warm and sunny (the first real heat we had felt since Omaha) that we found it hard to leave Phoenix. The road beckoned though, and we left the city of warmth. The desert offered no problems to the "300", and we shot through Aguila, and Quartzite. Then we saw the Colorado river and California. After being inspected by the border patrol, we continued on through more desert. At last we came out above Indio, Calif., and looked down on green pastures. We reached Beaumont in time for story telling and dinner. We stayed in Beaumont three days and rested for the run to San Francisco.

On December 6th we left for San Francisco. We followed Route 138 out of the Los Angeles area and at Lancaster we turned west again and hit 99 at Gorman. From there we shot down the Tehechapi Pass at 60 miles an hour. Then, suddenly, the whole San Joaquin Valley was spread out before us, and we were across the deserts for good, with the mighty Sierras on the right and part of the Coast Range on our left to guide us down the valley. At the bottom we had only 300 miles to go, which by then was only a drop in the bucket to us.

Highway 99 needs little description. It shoots like an arrow up the valley, no hills, and four lanes all the way. We passed through Fresno at dusk with only 200 miles to go. Soon all the Spanish names in the valley began to pass by us. First Madera, then Merced and its air base, then Modesto and finally Manteca where we were to leave 99 and take 50 west to San Francisco.

The Isetta was purring right along so we aimed for the bay, and in no time we were in the Coast Range. Then, as we came out of the mountains, the lights of Oakland became visible in the distance. A few moments later we were able to make out the bay bridge. Finally, on top of Telegraph Hill, our mission accomplished, we locked back on what we had done.

We had come a total of 2,643 miles, on approximately $22 worth of gas and an odd assortment of food and fun. We had carried a little over 500 pounds of luggage and our own weight on one cylinder, and had averaged 37.3 miles to the gallon of gas, while driving flat out the whole time. The car had not had one mechanical breakdown the entire time, and only needed everyday normal maintenance. Performance was exceptional during the whole trip, and I must say in closing, "Well done B.M.W.!"

WHY?
You don't have
to drive a
dinosaur!
GAS
You don't have to drive a "gas hog"
todays buying trend
60 MILES PER GAL!
BMW
Isetta
(PRONOUNCED I-SETTA)
'300'
COREY'S
Come See It!
We've Got It!
Call or drop in for a demonstration . .
CAR TOWN
2175 EMPIRE BLVD.
OPP. EMPIRE DRIVE-IN BU 8-8422
ONLY 1048.00 FULL PRICE
NO MONEY DOWN - BANK RATES 36 Mo. to pay

The right-hand-drive model is fitted with an orthodox bumper. The 54½in overall width makes for ease of traffic threading. With their 42-watt main filaments the headlamps provide an excellent light spread for night driving

ROAD IMPRESSIONS OF NEW MODELS

THE B.M.W. ISETTA

Right-hand-drive Version of All-weather Three-wheeler with Famous 295 c.c. Overhead-valve Single-cylinder Power Unit

EASE of entry and exit, better-than-average convenience in parking when space is limited, low running costs, the weather protection of a saloon car—the Isetta Plus three-wheeler with right-hand drive goes a long way to providing near-ideal transport for the commuter. But it offers more than that—much more.

The power unit is a modified version of one renowned throughout the motor-cycle world—the 245 c.c. B.M.W.—enlarged to 295 c.c., fitted with a 12-volt Noris starter-generator, and said to develop a mere 13 b.h.p. at 5,200 r.p.m.

The impression throughout was indeed one of driving a medium- rather than a high-performance vehicle. Acceleration is brisk but not snappy; there is modest torque at low engine speeds; and there is regular, faultless idling that implies smallish valves and low-lift cams. Overall there is an underlying suggestion of tough ruggedness promising many, many miles of running with freedom from petty tinkering.

Located on the right of the body, and driving the single rear wheel through a four-speed and reverse gear box and totally enclosed chain, the engine, cold or hot, would start at a touch of the ignition switch. It was a shade capricious in very cold weather for about a minute after the first start of the day (neither wanting the choke open nor closed) but thereafter the carburation was faultless.

At tickover speeds, noise heard inside the body was confined very largely to the regular beat of the exhaust. As the engine speed was stepped up (and particularly *while* it was being stepped up) the noise volume increased—a compound of induction and cooling-fan roar, mild drumming of the body panels and transmission wine.

But all noise is relative, and at its worst that inside the B.M.W. was considered to be loud only when the engine was peaking in the indirect ratios—at 15 m.p.h. in bottom, 28 m.p.h. in second and 40 m.p.h. in third. At the Isetta's usual cruising speed—between 45 and 50 m.p.h.—the noise level was insignificant, and seemed to come chiefly from the gear box. Maximum speed, by the way, was approaching 55 m.p.h.

Because of the engine's modest low-speed torque the gear change was pretty much in demand when the vehicle was being driven in busy city traffic. The lever is mounted on the right and protrudes through the body trim, convenient to the driver's right hand.

Initially the change was rather below the standard of what many would regard as desirable on a vehicle of the Isetta's type. Bottom gear occasionally required some fumbling when the vehicle was stationary. Second did not always mesh readily. Inspection showed that the linkage to the gear box had been assembled dry. Lubricating the clevis pins transformed the change. Thereafter all gears could be slipped into engagement quickly and lightly.

As would be expected with an engine-speed clutch, that on the Isetta took up the drive fairly quickly. However, the care needed to make smooth getaways in the first few miles soon became a simple, matter-of-course habit.

Clutch and brake pedals are located on

Left: Bird's-eye view of the cockpit, seen through the open sun-roof. The gear lever is on the right. Grouped round the speedometer are ignition-warning, flasher and headlamp telltale lights

Right: The engine is concealed by an easily removable body panel. Oil filler and dipstick are on the right of the Noris starter-generator, the fan blades of which are exposed

a common spindle, one on each side of the steering column. Both were excellently positioned for ease of operation. The 7in-diameter, hydraulically operated Girling brakes, by the way, were first class, required no adjustment during the test and needed only moderate pedal pressure even for experimental crash stops.

The driving position is very much sit-up-and-beg, and adequately comfortable as a result. Steering is through a worm-and-nut box. And, as on most cars with small-diameter wheels, the steering has little or no self-centring action and a rather "dead" feel. Nevertheless, there was no tendency to wandering on the straight and accuracy of line on corners was of a high order. Fast bend swinging produced next to no roll. There was little or no kick-back through the steering column. The turning circle, at 24 feet, could be reduced with advantage.

Suspension was firm and when the car was driven one-up on rough surfaces there was some tendency to pitching. Two- or three-up (the bench seat will accommodate two adults and an average 12-year-old) and with the luggage platform behind the seat laden, the ride was considerably improved. No discomfort was experienced on long journeys. A point of interest about the suspension characteristics, is that children usually troubled by travel sickness were virtually unaffected in the Isetta.

After-dark driving was a pleasure indeed. The 12-volt 7in-diameter headlamps provided all the light anybody could want and the steering-column-mounted dip-switch fell conveniently close to the left hand.

Other points pleased: the horn was powerful; the tubeless tyres lost no pressure during the Isetta's three-week stay in our hands; located behind the seat squab, the fuel tap could hardly be more easy to operate; the amount of luggage that could be stowed on the shelf is far more than seems possible when you start packing.

There is excellent all-round visibility, marred only by slightly too wide windscreen pillars and, in wet weather, by steaming up unless an anti-mist compound was used all round; the electrically operated screen wipers swept wide areas and were adequately fast for most conditions; the interior was free from draughts —yes, even when the roll-up part of the roof was open; the roof could be opened or closed from a normal driving position; and the heater (fitted as standard) took the chill off the interior.

How to sum up? The Isetta is not a teenager's vehicle. It is more a car for someone seeking transport rather than sport, whose aim is economy rather than luxury, but who is prepared to sacrifice nothing in terms of reliability. Overall fuel consumption worked out at 59 m.p.g. That involved daily use for a to-and-from Dorset House run of 16 miles each way in traffic so dense that the journey invariably took an hour. On out-of-town runs, when the car was one-up and cruising speed was in the region of 50 m.p.h., the figure went up to close on 70 m.p.g.— and who can ask for better than that?

Simplicity of entry and exit could hardly be greater. The door swings open automatically as soon as the handle is turned

SPECIFICATION ▶

ENGINE: B.M.W. 295 c.c. (72 x 73mm) overhead-valve single with fan cooling. Compression ratio 6.8 to 1. Wet sump lubrication. Oil capacity 3 pints.

CARBURETTOR: Bing 1/22.

IGNITION and LIGHTING: Coil ignition fed from Lucas 12-volt, 32-ampère-hour battery charged by Noris 130-watt starter-generator. Double-dipping 7in-diameter headlamps with 42/36-watt bulbs Steering-column-mounted dip-switch. Twin tail-and-brake-warning lamps with 6/21 watt bulbs Flashing indicators

TRANSMISSION: B.M.W. gear box in unit with engine, driving through single-plate dry clutch, transverse shaft and chain enclosed in an oilbath case. Four forward speeds and reverse. Gear ratios: bottom, 23.2 to 1; second, 12.14 to 1; third 8.17 to 1; top 6.1 to 1; reverse, 30 to 1.

TYRES: Dunlop tubeless; 4.80 x 10in on all three wheels.

BRAKES: Girling, 7in-diameter; twin-leading-shoe front single-leading-shoe rear. Hand-brake operates on rear wheel.

SUSPENSION: All three wheels carried on pivoted arms. Front controlled by coil springs and hydraulic shock absorbers; rear controlled by quarter-elliptic leaf spring and hydraulic damper.

FUEL CAPACITY: 2¾ gallons including 5 pints reserve.

DIMENSIONS: Track 47¼in; overall width, 54½in; wheelbase 58in; overall length, 90in; weight (fully equipped and with approximately one gallon of fuel), 815 lb.

PRICE: £359 19s 6d including British purchase tax. Spare wheel and tyre £7 0s 6d.

MANUFACTURERS: Isetta of Great Britain, Ltd., New England Street, Brighton, 1, Sussex.

B.M.W. ISETTA 300 CABRIOLET MUNICH GERMANY

1150	1	13	60	50	7' 6"	4' 6"	4' 4"	770

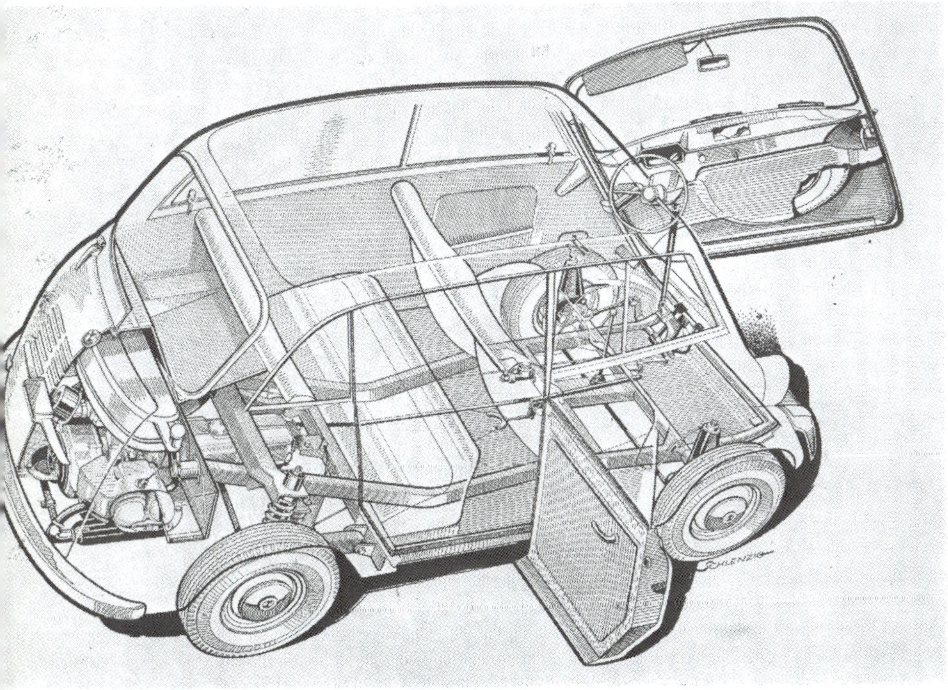

B.M.W. ISETTA 600 SEDAN MUNICH GERMANY

BMW

The BMW line of cars, manufactured in Germany, provides any prospective buyer with a very wide range in price power, and size.

Smallest and least expensive in the line is the BMW Isetta. This car is manufactured in two basic models. One is powered by a 15.3-cubic-inch, single-cylinder engine. The other is the Export model 57 powered by a slightly larger engine. The model 57 with some differences in body styling, bumpers, and lights is designed for the U.S. market and known as the BMW Isetta 300.

The model 57 and 300 are powered by a single-cylinder, four-cycle, air-cooled, ohv engine which has a 7-to-1 compression ratio and produces 13 bhp. The transmission is a four-speed box attached to the engine and the final drive is by enclosed chain to the closely spaced rear wheels. The narrow 20.4-inch rear tread eliminates the need for a differential. The body design is unique in that a single front door provides access and forms the entire front of the body. The steering wheel with its articulated-joint column and instrument cluster is anchored to this door. The Isettas are available in the standard model with a folding canvas sunroof, the Cabriolet with the rear-portion folding top and the Tropical model with ventilators in the front door. This little two-seater has a top speed of about 53 mph and fuel consumption is in the 50 to 60 mph range.

The BMW 600 is popularly called in the U.S. the Isetta 600. There are some differences in lights and bumpers on these cars depending on the countries in which they are sold. This car also has the single front-door access to the front seat, but being a four-seater, a second door is provided on the right side. The engine in the 600 is a detuned version of a BMW motorcycle engine. This powerplant is four-cycle, air-cooled and has two ohv horizontally opposed cylinders. The displacement is 35.4 cubic inches, compression ratio 6.5 to 1, and the output 26 bhp. The transmission has a floor shift and four synchromesh forward speeds. Unlike the 300, this car has conventional rear wheel spacing and a differential. The 600 can top 60 mph and fuel consumption is from 39 to 55 mpg.

Bayerische Motoren Werke, AG, Aktiengesellschaft, 76, Munchen 13, Germany. BMW 600 and Isetta 300 imported in U.S. by Fadex Commercial Corp., 487 Park Ave., New York 22, N.Y.; 319 South Van Norman Rd., Montebello, Calif. Other models distributed in U.S. by Hoffman Motor Car Co., 443 Park Ave., New York 22, N. Y.; Hoffman of California, Inc., 9130 Wilshire Blvd., Beverly Hills, Calif.

The BMW 300 is a tiny two-seater equipped with a single-cylinder engine developing a mere 15.5 bhp. Its only door is front-mounted.

The BMW 600, similar in styling to the 300, in addition to the front-mounted door has a side door for its back-seat passengers. It carries four people and is available as a hardtop or sunroof model.

BMW ISETTA - 300

ENGINE CAPACITY: 18.16 cu in, 297 cu cm;
FUEL CONSUMPTION: 72.3 m/imp gal, 60.3 m/US gal, 3.9 l x 100 km;
SEATS: 2; MAX SPEED: 52.8 mph, 85 km/h;
PRICE: list £ 268, total £ 325.

ENGINE: 4 stroke, on right; cylinders: 1, vertical; bore and stroke: 2.83 x 2.87 in, 72 x 73 mm; engine capacity: 18.16 cu in, 297 cu cm; compression ratio: 7 : 1; max power (DIN): 13 hp at 5200 rpm; max torque (DIN): 14 lb ft, 1.9 kgm at 4600 rpm; max number of engine rpm: 6000; specific power: 43.8 hp/l; cylinder block: light alloy; cylinder head: light alloy, hemispherical combustion chamber; crankshaft bearings: 2, on roller bearings; valves: 2 per cylinder, Vee-slanted at 80°, with push rods and rockers; camshaft: 1, side; lubrication: gear pump; lubricating system capacity: 1.5 imp qt, 1.8 US qt, 1.7 l; carburation: 1 Bing l/22 131 horizontal single barrel carburettor; fuel feed: gravity; cooling system: air-cooled.

TRANSMISSION: driving wheels: rear; clutch: single dry plate; gear box: mechanical; gears: 4 + reverse; gear box ratios: (I) 4.22, (II) 2.23, (III) 1.53, (IV) 1.16, (Rev) 5.25; final drive: chain transmission; ratio: 2.31 : 1.

CHASSIS: tubular; front suspension: independent, swinging trailing arms, coil springs, telescopic dampers; rear suspension: rigid axle, cantilever ¼-elliptic leafsprings, telescopic dampers.

STEERING: screw and nut; turns of steering wheel lock to lock: 2.50.

BRAKES: drum; braking surface: front 33.48 sq in, 216 sq cm, rear 16.74 sq in, 108 sq cm, total 50.22 sq in, 324 sq cm.

ELECTRICAL EQUIPMENT: voltage: 12 V; battery: 24 Ah; dynamo: 130 W; headlights: 2 front.

DIMENSIONS AND WEIGHT: wheel base: 59.05 in, 1500 mm; front track: 47.24 in, 1200 mm; rear track: 20.47 in, 520 mm; overall length: 92.72 in, 2355 mm; overall width: 54.33 in, 1380 mm; ground clearance: 5.51 in, 140 mm; dry weight: 816 lb, 370 kg; turning radius (between walls): 13 ft, 4 m; width of rims: 3''; tyres: 4.80 - 10; fuel tank capacity: 2.86 imp gal, 3.43 US gal, 13 l.

BODY: coupé; doors: 1 front; seats: 2.

PERFORMANCE: max speed in 1st gear: 12.4 mph, 20 km/h; max speed in 2nd gear: 24.9 mph, 40 km/h; max speed in 3rd gear: 37.3 mph, 60 km/h; max speed in 4th gear: 52.8 mph, 85 km/h; power-weight ratio: 62.8 lb/hp, 28.5 kg/hp; useful load: 507 lb, 230 kg; max gradient in 1st gear: 28 %; speed in top at 1000 rpm: 9.7 mph, 14 km/h.

PRACTICAL INSTRUCTIONS: fuel: petrol, 80-85 oct; engine sump oil: 1.5 imp qt, 1.8 US qt, 1.7 l, SAE 20 (winter) 40 (summer), change every 900 miles, 1500 km; gearbox oil: 0.5 imp qt, 0.6 US qt, 0.6 l, SAE 40, change every 7500 miles, 12000 km; final drive oil: 0.2 imp qt, 0.2 US qt, 0.2 l, SAE 40, change every 7500 miles, 12000 km; greasing: every 900 miles, 1500 km, 9 points; sparking plug type: 240°; tappet clearances: inlet 0.006 in, 0.15 mm; exhaust 0.008 in, 0.20 mm; valve timing: (inlet) opens 6° before tdc and closes 34° after bdc, (exhaust) opens 34° before bdc and closes 6° after tdc; tyre pressure (medium load): front 17 psi, 1.2 atm, rear 14 psi, 1.0 atm.

The BMW "Bubble"

by Virginia DeMoss

PHOTOGRAPHY BY DAVID GOOLEY

At a recent BMW rally in Monterey Park, California, owners and enthusiasts circled the large parking lot of a Bavarian restaurant, mentally rubbing their hands over the assembled exotica. Two official judges made the rounds, inspecting and conferring, but in addition, everyone who turned out carried a ballot for the people's choice award. If one stood in the right spot, it was pretty easy to tell just which way the votes were going. The *right spot* was somewhere in close proximity to an immaculately detailed BMW Isetta 300 Moto Coupe.

Isettas, the miniature, bubble-shaped, front-opening cars with tiny close-together rear wheels, remain unfamiliar to many BMW *aficionados*, possibly because they came late to the fold and didn't stay long. Initially designed and built by Iso Werk of Italy, maker of such costly automobiles as the Iso Milano and Rivolta, Isetta is a diminutive of Iso, meaning, essentially, "small Iso".

BMW bought license to the car and by 1955 began turning it out with some slightly modified lighting and trim, and powered by their R/27 motorcycle

engine. The motorcycle engine, putting out a mere 13 horsepower in the Isetta, was highly detuned to get better gas mileage, for greater reliability, and to allow it to pull around a much heavier vehicle of some 780 pounds. Other modifications included a starter/generator on the end of the crankshaft, a blower housing to cool the engine, points and ignition coil, slightly different carburetion, and a modified transmission that, of course, includes reverse. But although the Isetta's is a motorcycle engine, virtually no parts on the single-cylinder upright are interchangeable with the standard R/27 powerplant.

Many of the Isettas marketed overseas – particularly in Britain – were three-wheelers because such vehicles qualified as motorcycles and weren't subject to automobile licensing and road tax requirements. Additionally, the car was made in four basic models: bubble-window and sliding-window coupes, and bubble- and sliding-window convertibles. There was also, for a time, a funny-looking little pickup that is now quite rare. New, the bubble-window models cost about $995 to $1098, the sliding window version about $1165. Never hot sellers on this continent, however, Isettas were often discounted to as little as $800 and were sometimes even given away free with Cadillacs.

Isettas definitely have an interesting past, and any one of the unusual little cars is bound to cause a stir among car lovers, but this one, belonging to Marilyn Felling of Topanga Canyon, California, more so than most. The 1957 green Moto Coupe sports original paint, hand-rubbed out by Felling, and has been restored to perfect condition by her. When she found it it had been stored under a thick layer of dust and grime and was filled with garbage and rats' nests. She disassembled it down to the last nut and bolt – including the 100-odd pieces in the enclosed chain case – fixing, cleaning and replacing. Many of the pieces on the car are brand-new, having lain in their original BMW wrappers for more than 20 years. Felling had been, in fact, collecting and savings such items – and having others made – for nine years in anticipation of finding and restoring this particular model.

Ms. Felling, founder and president of the Heinkel, Messerschmitt, Isetta Club of America, and the largest and virtually only supplier of Isetta parts in the U.S., has a love for the little cars that goes all the way back to her childhood. In 1958, walking home from the third grade, Marilyn spotted one of the cars and "just fell in love with it. Being a little kid, I liked it because it was a little car, and I knew that when I grew up I was going to have one."

Ten years later her dream hadn't diminished, and so her husband, Carl Felling, in spite of his protests that the cars are unstable and dangerous, made her a gift of one on Valentine's Day. She immediately began trying to get parts for it. There was still one surviving dealership in Los Angeles

Isetta sunroof helps to dispel that claustrophobic feeling.

at the time (Marty's Foreign Motors, from whom Felling learned the parts business) that had parts and worked on Isettas, but it seemed that every single little thing Felling asked for brought the same reply: "It's no longer available."

Finally, frustrated, but still determined to return the object of her childhood fantasies to the road, she walked into Marty's one day and offered to buy out his entire Isetta stock. Marty agreed and Felling acquired a garage full of parts and pieces. She still didn't have everything she needed, but figured that turning her parts collection into a business was the best way to get in touch with those who did.

Felling met the head of BMW, who gave her enough clues to begin her search, and then painstakingly began making contacts herself, until she knew where to find everything still available. She also went back to all of the original secondary suppliers to BMW – including Aute, Boge, Bing – who were usually willing to make up parts again, provided the order was large enough. She went ahead and had everything reproduced that she could, which is another reason that much on her present car looks brand-new. Because of the quantity in which she had to have parts made in order to get a single one or two for her own Isetta, Felling had to drum up a lot of additional business to buy up the surplus pieces and cover the expense of reproduction.

Since then Felling has gone into the manufacture of other items (this year she's doing all of the rubber parts), and she and Carl frequently comb Germany and England to import many more. Since the cars were manufactured in five countries – England, Belgium, Germany, Italy, and Brazil – there are available parts scattered everywhere, most of which Felling knows about. . . . or will. Today she can get an Isetta owner anything he needs, provided he is willing to wait as long as three to four months for rarer pieces. Marilyn Felling is *the* source for Isetta parts these days, as witnessed by the fact that both BMW in North America and the factory in Germany refer all such requests to her.

Long a car enthusiast, and owner of some 20 different models, Felling's first instinct upon getting her first Isetta road-ready again was to join a club. As you might guess, however, such organizations were harder to find than parts for the little cars, and, in fact, didn't exist at all.

Such a minor detail was far from enough to stop Felling, however. If you can't find Isetta parts, you have them made; if you can't find an Isetta club, you start one. Seven years ago she did just that and began publishing the Heinkel, Messerschmitt, Isetta Club's

CLOCKWISE FROM TOP LEFT: Marion Felling, the bubble car booster and parts magnate. Isetta luxuriates on narrow, single lane roads. Note license plate. Profile view is probably car's best. The engine resides under the finned panel. VW styling is evident in this low-angle view.

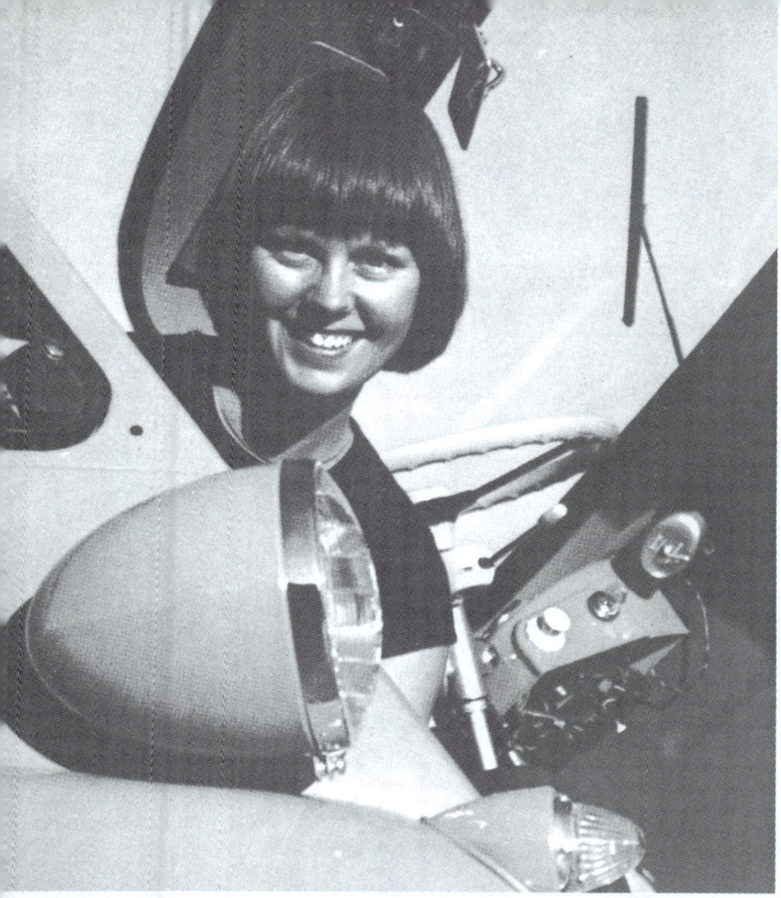

ISETTA

RETROSPECT

1957 BMW Isetta 300 Moto Coupe

Owned by Marilyn Felling

Photographed for *Motor Trend* by David Gooley

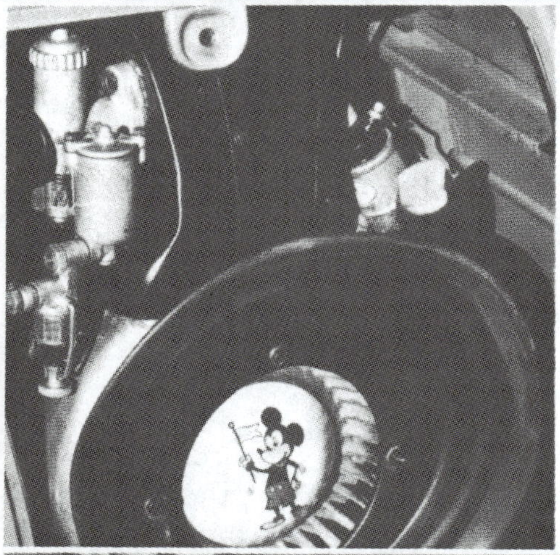

TOP: Isetta door allows unexcelled ease of entry. Note that steering column and instrument panel move, too. CENTER: Detuned BMW motorcycle engine is rated at an anemic 13hp. However, gas mileage compensates for lack of squirt. BOTTOM: The Isetta cockpit: as spartan as they get! Outstanding quality of restoration is obvious.

Continued from page 134

quarterly newsletter, *Bubble Notes*, which contains want ads, parts sources, reprints of original road tests and sales brochures, and technical and historical pieces on the fascinating little cars. The club currently has 250 members, 80% of which own Isetta 300s like Marilyn's, and the balance of which own Messerschmitts and BMW 600s, which are larger Isettas with a side door in addition to the front opening. Marilyn also owns one of the three-wheeled Messerschmitts and is one of very few who own a Heinkel, a rare car here in the States, that looks like a cross between an Isetta and a Messerschmitt.

The club has held a meet in various Eastern locales for several years now, and for the first time in 1977 also held a Western meet in Fullerton, California. This year, in addition to the California meet in June, and the St. Louis event, probably in September, the club is also planning an agenda of activities in conjunction with other car clubs, including the Pacific Austin Bantam Club's Annual Spring Tour and Picnic.

According to research conducted by Felling, approximately 8,000 Isettas were orginally imported to this country, about 1,000 of which are probably still in existence. Last year her newsletter published chassis numbers and owners names and addresses for some 425 U.S. cars; this year Felling hopes to publish at least 800. This is finally a time, according to Felling, when Isettas are coming into their own, partially because people are discovering that they're weird, neat little cars with some value to them, but most importantly because parts are once again available.

Isetta fanatics have Marilyn Felling to thank for that. According to her, "Parts just weren't available, so I made it my thing to see that they were because I love the cars so much. If I can make a part and get it for myself, then I can supply it to all those other people. It makes the cars run, it makes people happy, it makes me happy. I couldn't think of anything better to do to make a living than something like this that I enjoy so much."

ISETTA CAR?

Tony Marshall, newsletter editor of the Isetta Owners Club of Great Britain charts the life story of the famous Isetta, the 'bubble' that burst out all over the world

WAY BACK in days of yore, 1953 to be exact, the Italian Iso Company unveiled a vehicle that was to revolutionise personal transport for many thousands of people throughout the world. This car was the Isetta, the name meaning 'Little Iso', and together with later versions employing BMW mechanical components, it was to prove the best seller of all the odd-shaped little conveyances so accurately described by the expression 'Bubble-car'. True, the world had seen some strange vehicles in the past, many of them in what could loosely be termed the cycle-car class, but this was something different, and offered a degree of refinement that placed it in a class of its own.

The most striking feature was the shape. In length, the Isetta totalled only 7ft 4½in, about 4ft shorter than the smallest British car of the time. This reduction in overall length, compared with a conventional car, was the result of a completely new approach to the design, whereby the usual bonnet and boot areas were dispensed with altogether. The car was intended strictly as a two-seater, and the engine was situated at the rear, while luggage space was limited to a large shelf behind the seats. However, the most noticeable innovation about this startling new car was the door. Singular. There was only one. And it was at the front. In fact, it might be more accurate to say it was the front, for it accounted for almost the complete width, 40in out of a total of 52¾in. The other most obvious break with tradition involved the positioning of the rear wheels. Not unexpectedly they were at the back, but they were set very close together, and this gave the car a very odd appearance. The rear track was, in fact, only 19¼in, compared with a track at the front of 45¼in.

The Isetta was indeed a very strange car, and represented a completely new solution to a problem that had been nagging at light car designers since the birth of the motor car; how to manufacture a compact and economical runabout offering a reasonable standard of comfort and performance which was economical and would occupy the minimum of road space when in use, and parking space when at rest. Strange it may have been, but within a few years it became a common sight on the roads of Europe, and even in the company of high performance sports cars! This came about when, within less than a year of introducing the Isetta, the Iso Company entered a team of them in the gruelling Mille Miglia race of 1954. Five Isettas took part. All of them completed the course, and they took honourable positions in the Index of Performance. This was probably the most notable achievement of the Isetta in competition, although they were later entered in several international rallies, and usually put up a good showing in their class, although the competition must have been somewhat limited!

Peter Ustinov and early '55 Isetta from the film *Romanoff and Juliet*

However, only a small part of the Isetta's history took place in Italy. Like Henry Ford's Model T, the Isetta conquered large areas of the world. Also, like the Model T, the original models were offered only in one colour. In the case of the Ford, of course, it was black, but Iso only offered their baby in ivory. Nevertheless, it was possible to specify an alternative colour, provided the customer was prepared to accept a car assembled by the Belgian subsidiary company.

The Iso Isetta was never made available in Britain, but the BMW Isetta was on sale to British customers soon after the Bavarian firm commenced production in 1954. The German version of the Isetta was basically similar to the Italian, but sported a 247cc single cylinder four-stroke engine of BMW manufacture instead of the 236cc Iso two-stroke. This change boosted power to 12bhp, an increase of 2.5bhp.

In Germany the Isetta proved a great success, but this was not reflected in some of the other European markets where local taxation requirements favoured three-wheeled cars. Iso dropped production soon afterwards, and did not re-enter the field of motor car manufacture until 1962, by which time they had switched to rather more prestigious vehicles, and it is thought that this led to the cessation of assembly in Belgium. All the Italian factory plant was shipped to Brazil, where production of the original model was resumed, and continued until 1961 as the Romi Isetta. This, incidentally, created the curious situation that the initial model outlived all the later derivatives, the last of which disappeared at the end of 1964. The French Velam company was granted permission to build their own version of the Iso Isetta, but as all the equipment had gone to South America they had to start from scratch, setting up new tooling, and the Velam Isetta ended up with a rather different body.

In England the Isetta fell foul of the licensing regulations. As it had four wheels, it was deemed to be a normal car (normal??), and as such was taxed at the normal car rate. This proved to be a considerable barrier to sales, as such vehicles as the Messerschmitt and Britain's own Bond Minicar and Reliant were already established and were eligible for the reduced road fund licence at the tricycle rate. The concessionnaires, AFN Ltd, did all they could, but only a relatively small number of these early Isettas were imported into Britain. Happily, one of the very early ones used by AFN for demonstration purposes still survives.

An interesting variant appeared in Germany at about the same time as BMW started producing the Isetta. This was the Hoffman Auto-Kabine 250, and it appears to have been virtually a direct copy of the Isetta, but with a conventional door on the right hand side instead of the now familiar front entrance of the Isetta. In fact, the Hoffman was so similar to the Isetta that BMW instituted court proceedings for infringement of their patents, and Hoffman production was brought to a halt in February 1955 after a total of only 113 had been completed.

1957 saw the introduction of a re-styled Isetta, featuring sliding windows in place of the earlier fixed windows with swivelling quarter lights. This new model was generally fitted with an enlarged engine of 298cc, and was designated the 'Export' model in its native Germany to differentiate between it and the 'Standard' model which continued in production for a short time thereafter.

It was in 1957, also, that assembly of Isettas commenced in Britain, using a building that had formerly housed the locomotive works of the London, Brighton, and South Coast Railway, absorbed by the Southern Railway in 1922, and the Southern Region of British Railways in 1948. In this respect, the Isetta can compare with the Scootacar of 1957 to 1965, which was built in the workshops of the Hunslet Engineering Company of Leeds, another locomotive manufacturer. The use of the Brighton locomotive works by Isetta Limited, as the new company was called, led to the unlikely situation that, since there was no road access, even the workforce and visitors having to descend a flight of over 100 steps, the only way in for raw materials, and out for completed cars, was by rail. The British Isettas, which were all of the sliding window type, used many components manufactured in Britain, including Rubery Owen chassis, Girling braking equipment, and Lucas lights. The first completed car rolled of the assembly line on April 23, 1957, and between that date and the end of 1964, when production ceased, some 30,000 British Isettas were built, which sounds impressive until compared with a total German production of over 160,000!

Sales of the Isetta in Britain were never as high at this time as those of the rival bubble cars, because of the licensing laws that mitigated against four-wheeled vehicles, so in 1959 Isetta of Great Britain announced their new three-wheel model. This was the turning point as far as sales in Britain were concerned, and the Isetta soon became established as a common feature on the highways and byways of the nation. The four-wheeler model continued to be made alongside the newcomer, but mainly for export. (Isettas of British manufacture are still to be found in the Dominions of Canada, Australia, and New Zealand.) The final major new feature to appear was right-hand drive, which first became available towards the end of 1959, although there were forays into the field of specialist vehicle manufacture with the introduction of a two pedal version using the Smiths Selectromatic automatic clutch, and some small batch productions of van and pick-up types. There were even some Isettas specially made to be driven by disabled drivers, in which the steering wheel and other controls were replaced by a bar of a similar type to that still used on the current invalid tricycles. These disabled driver Isettas also featured the Smiths transmission, and special seats for ease of entry and exit.

Above: part of an original sales brochure. The dealer's name at the bottom reads 'H.W. Motors Ltd. George Abecassis . . .' Left: from this angle, an Isetta chassis looks quite complex!

The Isetta is, and always was intended strictly as a two-seater, and, of course, this limited its appeal to potential buyers. The powers that be at BMW were not slow to recognise this and announced in 1957 a new four seat car based on the Isetta, and in many ways of similar appearance. Some of the components were common to both cars, but in the main there was little scope for interchangeability. The name of the new car was, rather un-imaginatively, the BMW 600, and it received glowing reports at the hands of contemporary writers with the motoring press.

The concept of the BMW 600 arose from the familiar theme of providing the maximum of usable space in a vehicle occupying the minimum of external space, and in this it excelled. It had a front door, as on the Isetta, for access to the front seat, which accommodated two people in comfort, being luxuriously upholstered with a shaped squab, as well as a door on the right for access to the rear seat. Behind the rear seat was a deep well for luggage, and this could be enlarged for the carriage of bulky items by folding down the squab. The engine, developed from the flat twin unit already used on the BMW R60 motor-cycle, but detuned for even greater flexibility and smoothness, was located at the rear, and access to it was via the rear panel which was hinged just below the rear window. On this car the four wheels were relatively

Above: detailed cross-section of a typical Isetta. Its manufacturers called the car '. . . a new solution to the motoring problem'

Amazingly, Isettas were fairly competitive racers in the Fifties. Here a local entrant crosses the finish line of the 1955 Mille Miglia

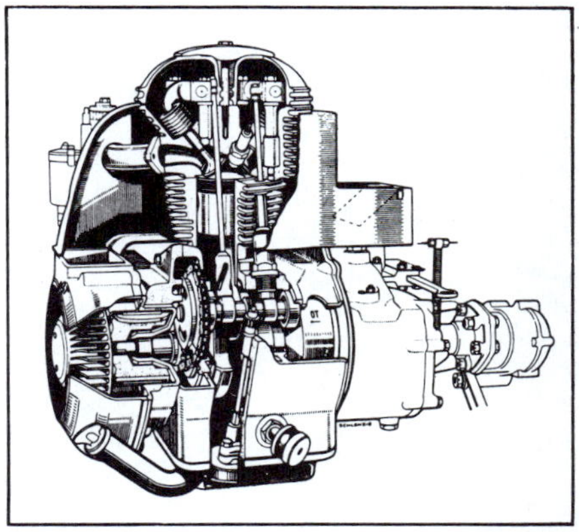

Two Isetta engine options were offered, 245cc and 298cc. Rated power was 12bhp and 13bhp respectively

evenly spaced, the front track measuring 49in, compared with a rear track of 46in. The whole car measured no more than 9ft 6in in length, yet experience has shown that the BMW 600 offered the sort of comfort normally associated with much larger and more expensive cars, and was quite suited to long distance touring. Not that the 600 was a cheap car. The price of those sold in Britain, thought to number only about 25, was £676 including tax in 1958, and this was quite expensive for such a small car when compared with the home produced products such as the Austin A35, Standard 8, and Ford Popular.

Production of the BMW 600 ceased in 1961, when it was replaced by the 700, which although using some of the same mechanical parts, was more of a conventional car, and would seem to be out of place in this story of the Isetta.

Direct BMW involvement with the Isetta ceased in 1962, although engines continued to be produced for the British factory until 1964, when production came to an end there, too. This left only the South Americans to carry on making Isettas, a situation which was to be repeated when the Volkswagen Beetle was dropped from the range at Wolfsburg, to be retained in production in South America.

For some years, the Isetta in all of its forms was only considered fit for use by the impoverished, and seldom made the headlines. An exception was the one which successfully smuggled 18 refugees from East to West Germany – not all at the same time! – concealed in a compartment behind the seat and above the engine, before being discovered. Recently they have become sought after by enthusiasts, and some near derelict hulks are being rebuilt with loving care for use both on the roads, and in exhibitions and shows. There are Clubs for Isetta owners in Germany, the United States, and in Great Britain, and these are proving invaluable to the restorer, as well as to the man who just wished to use his Isetta for every-day commuting, as sources of spare parts, encouragement, and information. It seems that the Isetta will be around for many years to come. □

This cutaway of the BMW 600 shows its side door, roomier body and extra seating. It was very much a 'step up' from the original Isetta

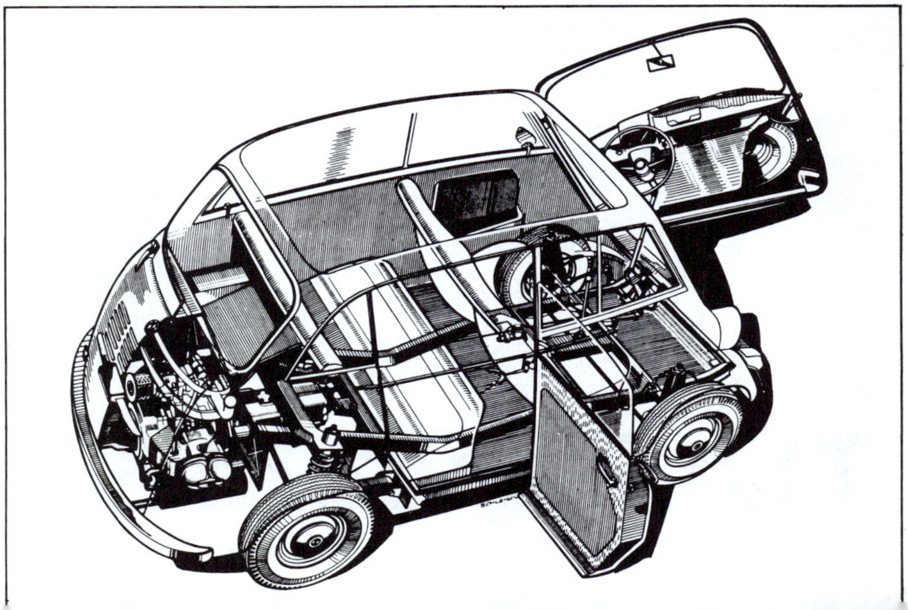

1957 BMW ISETTA 300

by Leonard Frank
photos by David Gooley

The Egg That Saved BMW

ONE MUST consider the time. Berlin was ruined, Germany divided. The oil fields in Rumania, factories in Poland,Czechoslovakia, Belgium, the Netherlands, Northern Italy, Great Britain—all in rubble. The Marshall Plan became the important fact in life.

England put VW back into business in their zone of Germany, and it was a success nearly from the beginning. Without taking anything away from the VW story, such was the hunger for transportation that any charcoal-burning Opel, any Ford Pilot V-8 converted to V-4, any old Renaults, Fiats, Austins, Citroens that could be persuaded to run came just after food in importance and usually well before shelter.

Not since the period surrounding World War I would so many half-baked cars hit the European market at once. The first time around it was ignorance (or innocence if you're kindly disposed), the second time, avarice. Crude two-strokes were in abundance, strange—or near absent—suspensions, wooden bodies, odd shapes, all sorts of weird configurations and amalgamations.

Bicycle manufacturers popped up where none had been. Old bicycle manufacturers began looking at scooters and motorcycles, motorcycle manufacturers at automobiles. It was no surprise, then, when Renzo Rivolta, who made Iso refrigerators in Milan, decided to go into the scooter, light motorcycle and three-wheel truck business. That was 1948. Success with these vehicles led him to look at cars.

The passenger-car design decided on was, well, unconventional. November 1954's *Motor Trend* asked coyly, "Do you like a few laughs?," and bid Rivolta's efforts adieu with, "Stay off streetcar tracks."

What Rivolta had designed and developed during 1952-54 was a box-section frame with tubular cross members. At the front was a variation of Andre Dubonnet's front suspension—trailing arms with horizontal coils worked by rockers. There was also a brief flirtation with rubber in compression. The rear was a bit more conventional—solid axle with trailing quarter-elliptics. Hydraulic shocks rear, friction front. Not too different.

But the dimensions were surely different: wheelbase, 58 inches; front tread, 47.2; rear, 20.4. Overall length was 89.9 inches—about four inches short of the wheelbase of a VW. Weight was less than half the Beetle's, displacement about one-fifth.

The thing that made it all go was a two-cylinder two-stroke with a single combustion chamber, one intake port and one exhaust port. The English Trojan of the twenties (itself a bit wide of the mainstream) and the German Zundapp motorcycle both used similar engines. In the Zundapp it was nicknamed a "twingle."

Transmission was a four-speed mounted in unit with the engine, and, like it, transversely. Drive from there to the differential-less rear axle was by chain enclosed in a light alloy housing, both swinging in unit.

But the thing most people saw and remembered was the egg-shaped body with front-mounted single door. The steering column was u-jointed and hinged, arranged so that it swung to make exit and entry to the narrow bench seat easy. The body was sheet steel over light tubing with plastic windows sides and rear on early cars.

The British, bless them, in an early road test in *The Motor*, pointed out that the door was similar to the arrangement on a Hansom cab. They went on to say that it "carries two people in style and comfort." With standards like this it's no wonder the English auto industry has so many present troubles. The venerable firm of Frazer-Nash rushed to import them.

The Isetta had a top speed around 45 mph and took near 36 seconds to reach 40. Remember that a contemporary VW did 0-60 in about 30 seconds. Fuel consumption in "hard driving" was about 50 mpg. It was really no competition for the latest Fiat Topolino 500C. Further, Fiat, Renault, Austin, Morris and

1957 BMW Isetta 300

Above and above right: *BMW's well-known blue and white roundel, which is supposed to symbolize a spinning aircraft propeller, appears on front opening door and on rear-mounted gas filler cap.* ***Right:*** *At speeds under 40 mph or so the Bimmer is quite stable.* ***Below:*** *Profile looks a bit like a truncated VW beetle. Use of egg shape for body adds strength to sheet metal without using weight-adding bracing.*

Citroen got back into volume production far faster than Rivolta had expected them to. In 1955, he got rid of the little Isetta, lock, stock and welding jigs, and moved toward the production of the Iso-Chevy hybrids for which he became best known.

However, before Rivolta divested himself of the econo-egg business, he granted a license to V.E.L.A.M. in France, Isetta Great Britain, Ltd., and Romi-Isetta in Brazil. Eggs also appear to have been assembled (hatched? laid?) in Belgium.

In 1955, BMW was busy building for the collectors and restorers of 1982. Certainly their line of baroque sedans attracted no attention in 1955. The 503 V-8 cabriolet and lovely 507 were overweight, overpriced, and in short supply even for the small demand. The costs of tooling for the sophisticated alloy V-8 and its chassis ate up scads of cash. BMW needed a new car to sell to the German mainstream, and needed it, as they say, for nothing down.

Germany was rebuilding rapidly—VW, GM (Opel), Ford Taunus, the Borgward group, Goggomobile (Glas), NSU-Fiat, Porsche and, of course, Mercedes, were cranking out cars and the Isetta seemed like unlikely company for them.

Then came Suez. Suddenly, little cars were "in," and the Isetta just the right size.

BMW immediately replaced the two-stroke "twingle" engine with a version of their R-27 motorcycle single. Whatever the theoretical advantage of the

Far left: *Rear of body is festooned with taillights and center-mount stoplight, tubular bumper guards and thin bumper.* ***Left:*** *BMW antedated the Mini-Minor in its use of 10-inch wheels on a passenger car.* ***Below:*** *Driver has to manipulate feet around u-jointed steering column to operate clutch and brake.* ***Bottom:*** *Deeply inset rear wheel position gives car a somewhat awkward stance.*

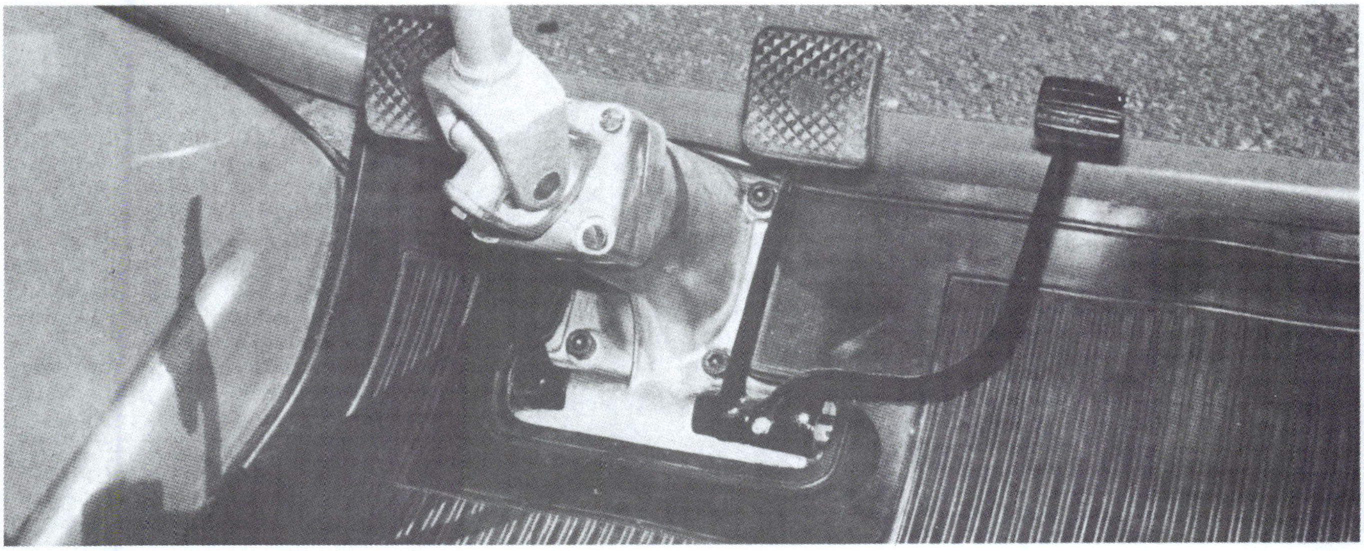

two-stroke, the four-stroke hemi-single was a big improvement. Both acceleration and gas mileage improved arithmetically—reliability, geometrically.

Headlights were raised from fender nacelles, a few functional and cosmetic touches were added, finish vastly improved, front suspension revised with vertical coils, and Iso became BMW. No parts interchange with the original.

For a few years, production and sales were good and enough cash was raised to allow development of the BMW 600. A kind of two-door Isetta, the 600 was the first BMW to use a version of the present rear suspension. The 600 was the genesis of the 700. The 700, with the exception of its rear-mounted opposed twin had most all of the elements of the 1500/1600/1800—first of the New Class cars that are the basis for all of today's BMWs.

Right: Inside the egg's shell. Steering column tips forward with door when it is open for convenient entry and exit. A tug on the steering wheel latches the door once you're inside. *Below:* Chain/rear axle casings are very clean castings. *Bottom:* Pushrod hemi single is mounted on right side of car forward of rear wheel. It pumps out 13 screaming horsepower. *Bottom right:* Ventilation comes from flip-out quarter windows and sliding fabric sunroof. Big side windows are fixed in place.

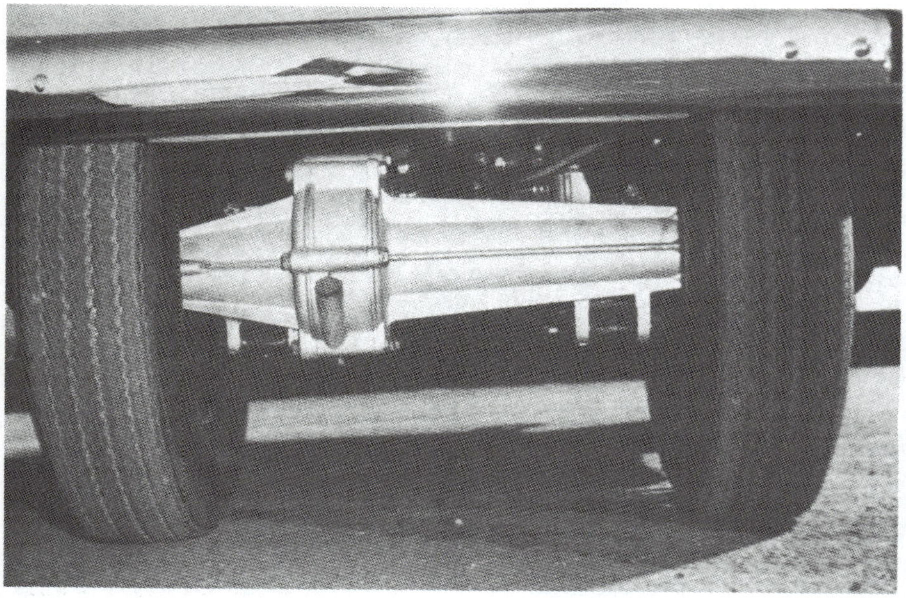

1957 BMW Isetta 300

By 1959 there were four variations in the Isetta's passenger bodies (sliding and non-sliding window coupes with and without sunroof) as well as pickup versions. In Europe, a 250cc single was available as well as the exported 300cc model. Competitors hatched the tandem seat Messerschmitt, the similar Heinkel (later the English Trojan), British Scootacar, the bizarre Dornier, Zundapp Janus....

By 1960, Suez was largely forgotten, VW had become a modern phenomenon and Alec Issigonis and British Motor Corporation had brought us to the modern small car with the sensational Mini-Minor. The day of the egg was past.

Here in the United States, we had taken our time taking the VW seriously. We would *never* take the Isetta (or Maico 500, Lloyd, BMW 600, 700, NSU Prinz, Fiat 500, Goggomobiles, or latter day Subaru 360) seriously. They were given away like oversized crackerjack toys to encourage the sale of normal cars. One Los Angeles Plymouth/Chrysler dealer would take a sledge to an Isetta on his TV commercials declaring he would destroy them rather than sell them any cheaper. There were plenty of Isetta dealers but nothing resembling a regular service network. The poor egg became a throwaway.

And in California, the final insult: the Isetta wasn't even freeway legal. What could possibly be more serious than that?

Driving Impressions

Nuts-and-bolts aside, there are debts to pay. Much of the information (and

specifications

Illustrations by Russell von Sauers, The Graphic Automobile Studio

© copyright 1982 Special Interest Autos

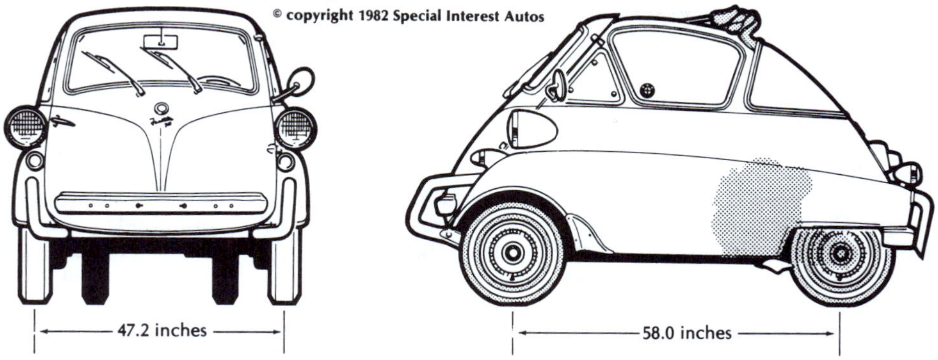

1957 BMW Isetta 300

Price	$998 plus accessories (approximately $50)
Standard equipment	Sunroof, heater.
Optional equipment	Mud flaps and fender guards, wheel trim rings, sun visors, steering and shift boots, whitewall tires, exhaust tip, pedal pads, passenger assist grip.
ENGINE	
Type	Pushrod (Hemi) air-cooled single.
Bore & stroke	2.3 inches x 2.9 inches.
Displacement	295cc (18.6 cubic inches).
Max. bhp @ rpm	13 @ 5200.
Max. torque	14 @ 4200.
Compression ratio	6.8:1.
Induction system	Single Bing carburetor, 22mm throat, gravity fuel supply.
Exhaust system	Single muffler.
Electrical system	Battery/dynamo (Noris) 12-volt.
CLUTCH	
Type	Single dry plate.
Disc diameter	5.75 inches.
Actuation	Cable.
TRANSMISSION	
Type	Four-speed constant mesh with reverse.
Ratios: 1st	10.95:1.
2nd	5.17:1.
3rd	3.54:1.
4th	2.70:1.
Rev	12.15:1.
DIFFERENTIAL	
Type	None.
Ratio	2.31:1.
Drive axles	Via chain and sprockets.
STEERING	
Type	Worm and nut.
Turns lock to lock	2.25.
Ratio	16:1.
Turn circle	24 feet.
BRAKES	
Type	BMW/ATE drums.
Drum diameter	7 inches.
Total swept area	49.9 square inches.
CHASSIS & BODY	
Frame	Welded tubes.
Body construction	Pressed steel over small diameter tube.
Body style	Single door, two-passenger coupe, fabric sunroof.
SUSPENSION	
Front	Trailing arms, coil springs, friction-type shocks on early cars, hydraulic on late.
Rear	Trailing chain case acting as rear-axle locator, quarter elliptic springs, hydraulic shocks.
Tires	4.80x10 bias-ply.
Wheels	Steel disc, 3.0x10.
WEIGHTS & MEASURES	
Wheelbase	58 inches.
Overall length	89.9 inches.
Overall height	52.6 inches.
Overall width	54.3 inches.
Front tread	47.2 inches.
Rear tread	20.4 inches.
Ground clearance	n/a.
Curb weight	Approximately 800 pounds.
CAPACITIES	
Crankcase	3.6 pints.
Cooling system	Air.
Fuel tank	3.4 gallons plus .8 reserve.
FUEL CONSUMPTION	
Av. city driving	Approximately 40-45 mpg.
Av. open road driving	Round trip L.A./Phoenix, AZ (average speed 47.5 mph), 51.7 mpg.
Marque Club Address	Heinkel, Messerschmitt, Isetta Club of America PO Box 90 Topanga, CA

most of the enthusiasm) for this article, as well as the car used for photographs, comes from Marilyn Felling, a pretty and otherwise normal lady who co-founded and pushes the Heinkel-Messerschmitt-Isetta Club of America.

Some are indifferent to cars, some like them, some are dotty about them. Mrs. Felling mainlines them. Especially the odd, the orphaned, the, well, interesting....

And I had to promise her that I would treat the Isetta seriously, not poke fun, give it a serious historical/technical treatment. OK, Marilyn. No smile on my face.

Since the Dynastart works directly on the crank, at crank speed, battery willing, starting is instant. The air (fan) cooled single warms quickly, the left hand shift lever has a very short throw and notchy feel, throttle is direct, the

Little BMW has good handling characteristics with minimal body lean. Tradeoff for good handling is choppy ride.

1957 BMW Isetta 300

engine clearly felt, steering quick with low effort. Brakes take a healthy push. If it weren't for the upright seating, fishbowl cockpit, and absolute lack of performance, you might think you were in a modern racing car. Come to think of it, the Isetta did win its class in the Mille Miglia.

Handling is surprisingly good. Body lean is nearly non-existent, there is no rubbery feeling in suspension or controls, no slack. The short wheelbase aids quick direction changes, the engine *sounds* torquey. Ride borders on harsh —ten inch wheels don't help (the genius of the Mini-Minor is ever more apparent)—but on a fair surface it becomes merely choppy. Perhaps we are spoiled. Noise level is high and on a hot day the fishbowl becomes a hot house. Directional stability in crosswinds, on bumpy roads, or on slippery surfaces is a problem. Still, the most common reaction to the Isetta is a smile, and that can't be all bad.

The shift pattern is a backwards H—takes a bit of getting used to. This early car has friction shocks that make a friendly chirp-chirp when they work.

Once in fourth gear the high noise level seems to recede. Or maybe one just gets used to it. Driving down the empty winding road there is no feeling of inadequate power, at least on this road with no one to pass and no real hills to climb. Handling really is race car-like—in slow motion.

The greenhouse effect can be countered by locking the sunroof open to ram air into the car. Noise goes up but one gets used to that too. Downshifts are a good excuse to double-clutch. The engine revs a bit slowly with heavy flywheel effect, but with practice, smooth, solid changes are possible.

Few cars have been built in five countries on two continents and still been considered curiosities rather than a cackling success. BMW produced the greatest number, about 160,000, between April 1955 and 1962. After Frazer-Nash (importers of BMW, Porsche and manufacturers of their own exotic car) stopped importing Isettas, control passed to Isetta G.B., Ltd., who eventually began to manufacture their own, including a three-wheeler to take advantage of motorcycle status—about 20,000 of all types produced between 1957 and 1963. The last new Isetta was produced by Romi, in Brazil, in 1965.

Unlike the Messerschmitt, the Isetta,

***Above:** Seating is strictly a two-person proposition. Body design gives you the feeling of being seated in a fishbowl. **Above right:** Shift lever "falls easily to hand." Becoming accustomed to backwards-H pattern of shifting might not be so easy. Emergency brake sits next to wheel well. **Right:** The little engine that could. Modified from a BMW motorcycle one-lunger, it will propel the little car to a top end of around 60 mph.*

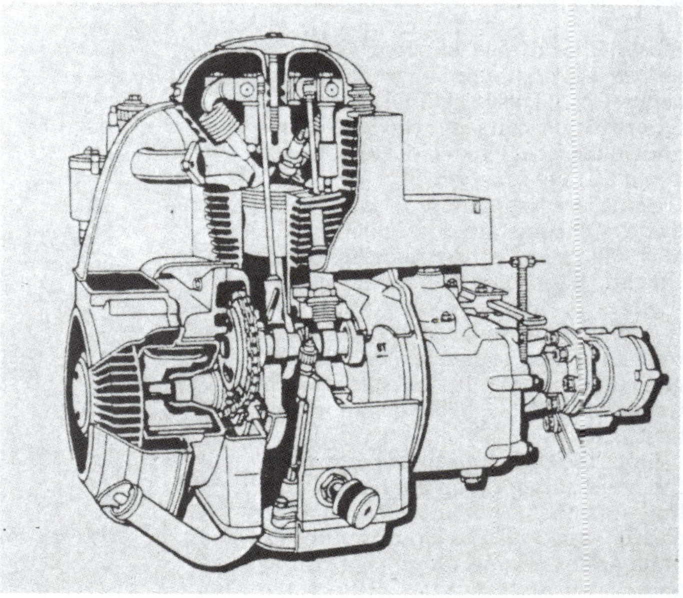

at least at low speeds, begins to feel like a normal car. The 'schmitt KR-175/200 is a sort of enclosed three-wheeled motorscooter that never lets the driver forget what it is. The Isetta is a clever, unorthodox variation of conventional design that deserved more development. A pity it didn't get it. □

Acknowledgements and Bibliography
Auto Age, *May 1955;* The Autocar, *Sept. 5, 1957; BMW Isetta 300 Accessory Index; BMW Isetta 300 Flat Rate Manual;* Bubble Notes, *various issues, a publication of the Heinkel-Messerschmidt-Isetta Club;* The Motor, *road test #30, 1958;* Motor Trend, *November 1954;* Road & Track *1959 Road Test Annual.*

Thanks to Jane Barratt, Librarian, Petersen Publications, Los Angeles, California. Very special thanks to Marilyn Felling, Topanga, California.

About the Author

Len Frank seems to have been born with a steering wheel in his hands. His twenty years of automotive experience include the ownership of a new car agency, new and used whole and retail car sales, domestic and import service work, retail and wholesale parts dealer, the marketing and production of speed equipment plus having a go in road racing in such diverse wheels as MG-TD, Berkeley, Saab, Ford Anglia, Porsche Speedster, Sunbeam Alpine and Fiat 8V Zagato coupe.

His automotive articles have appeared for many years in *Motor Trend, Sports Car Graphic, Cars, BMW Journal, Road Test,* where he was Deputy Editor, *Automotive Age, Car Classics* and other automags. He can be heard by Los Angeles readers as co-host of The Car Show on KPFK-FM. This is his first appearance in the pages of *SIA.*

Above: Spare hides behind seat. Big parcel shelf is handy for carrying pizzas. *Left:* The little cars even found their way into ads during the '50s. But wearing driving gloves to pilot a BMW Isetta is a bit much. *Bottom:* Rear styling has much more of a "scooter car" appearance than the front and sides.

ISETTA
OWNERS
CLUB
OF
GREAT BRITAIN

THE ISETTA OWNERS CLUB OF GREAT BRITAIN.
111, HERONSCROFT
COVINGHAM
SWINDON
WILTS SN3 5AW

SLIDING-WINDOW CONVERTIBLE
(EUROPEAN MODEL)

SLIDING-WINDOW PICK-
(EUROPEAN MODEL)

SLIDING-WINDOW COUPE
DELUXE (U.S. MODEL)

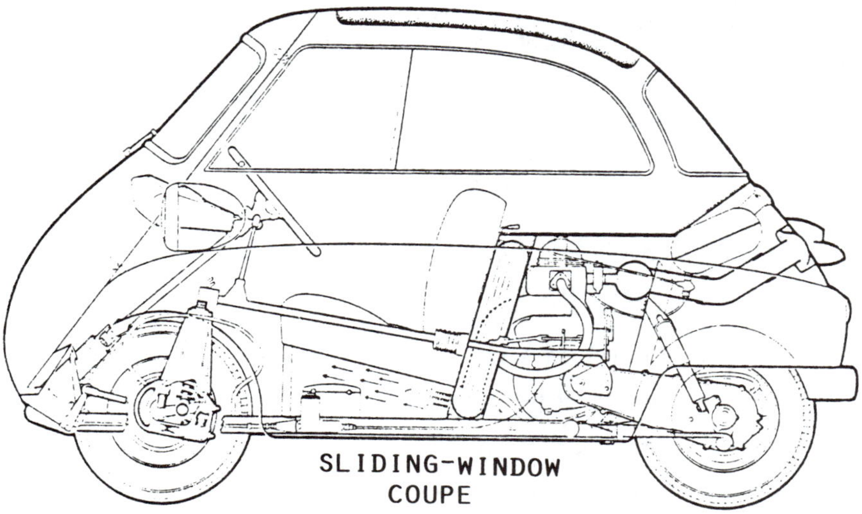

SLIDING-WINDOW
COUPE

BUBI

BUBBLE-WINDOW PICK-UP
(EUROPEAN MODEL)

tta

er's

de

BUBBLE-WINDOW COUPE
DELUXE (EUROPEAN MODEL)

NDOW

BUBBLE-WINDOW CONVERTIBLE
(EUROPEAN MODEL)

'The Isetta Lady'

Something new has been added to the small car inventory in Topanga, CA. Marilyn is shown with a 1956 Aerocar, which is to be restored.

Marilyn Felling has to be the kind of female every male car collector dreams about.

Not only is she charming and beautiful, but her entire life is devoted to the hobby!

In her own words, she's "blue-eyed, left handed, prefers semi-sweet chocolate to milk chocolate, and little cars are my whole life."

The charming Topanga, CA, woman does her own restorations, including mechanical work. Her business is manufacturing car parts and dealing in them.

She is also co-founder of a successful national club, and is both president and editor-publisher of the club's award-winning publication.

The cars the 32-year-old Marilyn collects are as unique as the lady herself.

In southern California she is known as "The Isetta Lady." Both in this country and abroad she is widely recognized as one of the leading exponents of the European-built miniature automobiles of the fifties (the cars some people call "cabin scooters").

Marilyn is co-founder of the fast growing Heinkel-Messerschmitt-Isetta Club of America, the HMI club for short. Her specific interest and expertise is the Isetta, but she is a prime promoter for all three makes.

How did this young, attractive and personable woman ever get involved in such small and obscure automobiles?

It all makes a lot of sense when you hear it from Marilyn.

"Back in 1958, when I was in the third grade, I used to walk home from school past an Isetta," Marilyn recalls. "I thought that when I grew up that would be the car I would own. But I didn't see another until I was 17. That renewed my interest.

"At that point I had met my husband, Carl, and we had together developed an interest in sports cars."

For some time Marilyn looked high and low for an Isetta, with absolutely no results, much to her fiancee's delight.

"Finally, I saw an ad for a 1957 Isetta for $200," she says gleefully. "I told Carl I'd quit eating, I'd go to work, I'd do anything on earth to have that car.

"He finally relented and said, 'Okay, this is your happy birthday, your merry Christmas, your happy Valentine's day.' He bought it for me on Valentine's day."

The purchase of the Isetta in 1969 soon led to the founding of the club.

"There was a letter to the editor in *Road & Track* inquiring if anyone was interested in forming a Heinkel-Messerschmitt-Isetta Club," says Marilyn.

"I responded to the letter written by Robert Maselko of Union, NJ. Come to find out he was a 13-year-old kid, and here I was a 21-year-old female, and between the two of us, 3,000 miles apart, we co-founded the club.

"We started out with a lot more enthusiam than anything else and obtained 40 members the first year through *Road & Track*. We now have more than 600 members, and I'm looking for up to 800 members this year due to such a growing interest in the cars."

Marilyn's business in Isetta parts began through the rapid growth of the club and the problems most members had in locating parts since most dealers had long gone out of business.

"I couldn't get parts for my first Isetta," says Marilyn. "But at that time there was still one dealership in the Los Angeles area, Marty's Foreign Motors in Torrance. I bought all their stock for $300. But what you get when you buy a dealer's old stock are all the parts that didn't sell well in the first place.

"In the next couple of years I bought out 10 more dealerships, and still kept getting the same crummy parts. Then I got on to importing from Germany. In about 1978, I finally bought all the remaining parts from BMW and they still recognize me as their supply source.

"So, it went from a very small part time business to a full time business that he just snow balled, considering what I'm selling."

However, Marilyn confesses that most of the parts whe sells are not NOS, but are new replacement parts manufactured by her or other members of the club. Members communicate quite openly, and when one members makes a part he usually has a large number made up which then become available to others through "Felling Enterprises."

Marilyn sells primarily Isetta parts, though she handles some Messerschmitt parts. Her business in Heinkel parts has to be pretty

What about these little cars Marilyn likes?

What are these cars – the Heinkel, the Isetta and the Messerschmitt?

The Messerschmitt is a three wheeled vehicle with an air-cooled two-stroke one-cylinder engine. It is considered by many historians to be one of the most unique and innovative vehicles ever produced in any quantity. More than 40,000 of the last model, the KR-200, were produced before production stopped in 1963. Only 300 Messerschmitts are known to survive in the U.S.

The Isetta was originally built in Italy by Iso, and in France and Belgium. In 1955, BMW bought the license and began building greatly updated series 300 and 600 models until 1962. About 8,000 were brought into this country, an estimated 1,000 survive.

The Heinkel was designed and built by Ernest Heinkel of Stuttgart, Germany. It has a three stroke, one cylinder engine, is air cooled and came with three or four wheels.

"The Heinkel looks like the Messerschmitt and the Isetta had a baby," laughs Marilyn Felling.

The Heinkel was built in Germany, then Ireland and finally England until 1965. Very few were imported.

How about an Isetta dune buggy?

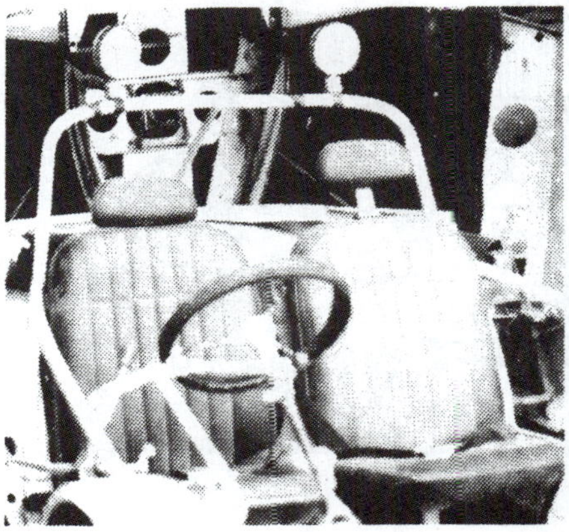

minimal since there are only 12 Heinkels in the club.

"I'm not selling Model T parts," kids Marilyn.

"I do not have Rolls-Royce prices. These are probably the cheapest cars you could restore today, and that's the way we intend to keep them. Our parts prices are about on par with Volkswagen and I probably have about 75 percent of the parts readily available."

Marilyn's personal collection consists of about a dozen cars including her original Isetta, a prize-winning 1957 Isetta, another all original Isetta and a 1956 Messerschmitt stored in a clear plastic bag for 15 years.

According to Marilyn the Messerschmitts are the most expensive of the three, with prices for prime examples in excess of $5,000. A good Isetta can be purchased for half that amount or less.

As interest in the little cars has grown, Marilyn's business and club activities have become fully computerized. She owns both a home computer and a word processor.

"When I graduated from art, I never dreamed I would be restoring cars and dealing in parts and running a car club," she told the *Coast Car Collector*.

"But it's a nice feeling to know that you've done something to preserve and restore a breed that would have disappeared if nobody had done it. Why do I do it? Well, somebody had to, and they chose me." – *Tim Howley*

At top right Marilyn Felling demonstrates that she's ready to take on any restoration job. Here, she's shown with a rough, but solid and complete 1956 Messerschmitt. At right are two photos of a 1958 BMW 600, owned by Ron Engel of Davenport, IA. Marilyn is the president of the Heinkel-Messerschmitt-Isetta Club.

The BMW 2002 cabriolet saloon ti,tii touring turbo

a comprehensive guide to the classic sporting saloon

BMW 2002 A Comprehensive Guide. Written especially for Brooklands Books by respected motoring journalist James Taylor and well known marque specialist Mike Macartney, this tite is a must for all 2002 enthusiasts. Models covered: Cabriolet, Saloon, ti, tii, Touring and Turbo. 176 pages, 20 page of colour plus 120 black and white illustrations. Hard bound.

ISBN: 9781855203341

From specialist booksellers or, in case of difficulty, direct from the distributors:
Brooklands Books Ltd., P.O. Box 146, Cobham, Surrey, KT11 1LG, England Phone: 01932 865051
E-mail us at info@brooklands-books.com or visit our website www.brooklands-books.com
Brooklands Books Australia, 3/37-39 Green Street, Banksmeadow, NSW 2019, Australia. Phone: 2 9695 7055
CarTech, 39966 Grand Avenue, North Branch, MN 55056, USA Phone: 800 551 4754 & 651 277 1200
Motorbooks International, P.O. Box 1, Osceola, Wisconsin 54020, USA Phone: 800 826 6600 & 715 294 3345

I've Been Hooked Since 1958...

An Original Bubble Nut

by
Barry W. Collins
Demopolis, AL

Those of us who are past or passing the "BIG Five-Oh" never think of ourselves as **old.** Heaven forbid! It's only when something like the comment in a recent **Bubble Notes** about Isetta "original users" comes to the fore that approaching senility rears its ugly head. I never really thought about the fact that the large majority of the HMI members never experienced the joys of original ownership of their little cars. Since I bought two of them off the showroom floor, so to speak, I will be happy to share as best I can the feeling of being one of the **original** Bubble Nuts!

First of all, neither of them came off of a showroom floor, but that's getting a little ahead of the story. The first time I saw one of these strange apparitions was the spring of 1958. The family, consisting at that time of a wife and two kids, was driving past the parking lot of a local supermarket and there it was. . . a bright green Isetta 300 with sliding windows. It was love at first sight. Even though we were travelling at the time in what was to become a classic on its own, one of the original 1955 Chevrolet V-8's, I would have swapped **even** at that moment! We stopped and gave the little bug a thorough look. The owner was apparently just passing through Demopolis and was inside the market shopping. I never did meet him (or **her,** Marilyn) or even find out his name. Regardless, the die had been cast.

"An I-what?"

Somehow, one of these little cars had to join the Collins family. I figured when we made a trip to a town with a BMW dealer, I would drop by and discuss terms. I had no way of knowing that even that early, BMW was beginning to get hoarse from saying, "An I-what?".

Demopolis is a small West Central Alabama town about 150 miles from the Gulf Coast. Later in 1958, the family headed south for a vacation somewhere in north Florida. After a pleasant few days, a tired Collins crew was back as far as Mobile, headed toward Demopolis. All of a sudden, 4-year old Marie shouted, "Daddy, daddy, there are a bunch of those little cars!" A screech of the brakes and a quick left turn got us back around the block to the Big Three Motor Company. And sure enough, there were a "bunch" of those little cars. It looked for all the world like a large Easter rabbit had laid the biggest nest ever of eggs. The Big Three Motor's lot was literally covered with all colors of sliding-window coupe Isettas ever built. Thinking back on that scene, there might have been a hundred or more of the cars. From all appearances, on that particular Friday afternoon, business did not appear to be booming. In fact, when we stopped, we seemed to be the only prospect in quite a while, judging from the reception we got. The four of us were standing on the lot ogling the shining little beauties when a smiling salesman appeared from the small building adjoining the lot. After the usual exchange of amenities, he asked if I was looking for a second car. I told him, no, I was interested in trading my Chevrolet for one. I had no intention of doing this, but thought it was as good a place as any to start price negotiations. He said the price of the Isetta was $1275, not including tax. After much discussion and haggling, he had finally talked me into keeping my car and buying the Isetta outright. Now we were getting somewhere. He was telling me what to do! He suggested we take it for a drive around the block to see how it drove. Rosie and I pounced onto the seat through that gaping front door and left kids in the care of the total stranger who was trying to relieve me of as much of my money as possible.

"He said the price of the Isetta was $1275, not including tax."

All the cars I had ever owned up to that point had either had automatic transmissions or column shift. In college, my roommate and I did have a floor shift '37 Chevy coupe. . .but what is this backward shifting handle sticking out of the left wall of this little car? Thankfully, there is a decal on the fender well which gives a pretty good clue as to the shifting pattern. We bounce off the lot and make our way out into the street. Wow! This is like **nothing** before. The pickup is amazingly good, the clutch works. . .we stop at the corner for a stop light. The brakes are a little spongy, but how bad a problem can that be with a 780 pound car? We really push our luck and go around not one, but **two** blocks, all the while grinning broadly at all the folks grinning at us. The "wall-shift" is mastered, we are hooked. . .all the friendly baby sitting salesman has to do is reel us in.

When we bounded back onto the lot, he came over to assist us out. He asked if I was ready to buy the car. I told him I was interested but the price was way out of line. (What could it hurt to say this?) He said, "Well, what will you give me for one?" I said, with as straight a face as I could muster, "Seven hundred and fifty dollars". He did the usual salesman's frown and said, "Mr. Collins, I can't do that, but I'll tell you what I can do. Yesterday a bicycle ran into that little red one on the corner of the lot up there, and if you don't mind the small scratch, you can take it home for 850." We went and looked at the car with the "collision damage" and almost laughed when we saw it. There was a tiny scratch which had already been filled in with paint and was for all practical purposes, invisible. I told him this was OK, but we didn't have a way to get it back to Demopolis. Even in the heat of the enthusiasm of becoming an Isetta owner, neither Rosie nor I, cherished the thought of driving the little bug 150 miles home that night. It was at this point we realized just how badly this gentleman wanted to sell an Isetta. His instant response was, "OK, I'll tow it to Demopolis tomorrow for you." I gave him a check and we were on our way back home to await the new family member.

". . . neither Rosie nor I cherished the thought of driving the little bug 150 miles home that night."

The next afternoon a big black Buick with a little red Isetta pulled into our driveway. The towbar being used was a standard option of the time and allowed the Isetta to be towed with the "big" car to wherever the family was going. We took delivery and began what I planned to be a lifelong relationship with the Isetta. The "Redbug" quickly became a familiar sight around town as it was used everyday for transportation to work or any other short

"The brakes are a little spongy, but how bad a problem can that be with a 780 pound car?"

An Original Bubble Nut

"It was accepted as an early example of the ultimate economy car."

trip driving. It was accepted as an early example of the ultimate economy car. The Isetta, while not a great performer by present fuel economy standards, performed in 1958 and many years thereafter like no other little car around. Realistic gas mileage was then pretty much like it is now. With normal driving around town, 32 to 38 MPG; on the road, much better if under 50 MPH.

Not long after we took delivery of that wonderful little car, an Encyclopedia Britannica salesman appeared at our home one evening. Shortly after answering the door we discovered he could care less about selling us encyclopedias. He was at our house because he saw an Isetta in the driveway! Besides, he was driving an Isetta. We welcomed him in as we would any Isetta owner and for the next four or five hours heard some of the most amazing stories imaginable about owning and maintaining an Isetta. The young man was from Canada. He had driven his Isetta from Canada to Demopolis and was heading for the west coast. He gave us a blow by blow description of carrying and **installing** a spare engine on the road. Looking inside his car was a little like opening the door of a well equipped Isetta parts house. He was ready for all contingencies. Of the many stories he related, one sticks in my mind above all others. He said in his home town he regularly performed at the half-time of local football games. The performance consisted of taking off from the end zone in his Isetta and upon reaching the 15 or 20 yard line he would put on the brakes and put the steering wheel into a sharp left turn. With the usually wet turf of the field, he said he could get a minimum of three complete 360 degree spins out of this. He would do this all the way from one end of the field to the other! A derivation of this procedure was to park the Isetta in a slanted parking place downtown by doing a similar spin of 180 degrees so the front of the car pointed out from the curb. Since I must have looked a little skeptical about these claims, he said, "Come here, let me show you". With that he lept into his car, drove out onto the street in front of our house and came screaming across the side yard doing a series of sideways wheelies all the way to our hedge. He definitely did what he said he could do. I've wondered many times if he ever got to California, if he ever sold any encyclopedias, or if he might still be doing spins across the varying terrain between here and there.

I discovered an interesting fact about my daughter in the early years of Isetta ownership. Even though she was the one who spotted the "bunch" on the Big Three lot and loved the little car like the rest of us, changes began when she was 7 or 8 years old. Frequently I would pick her up at one of her friend's homes on my way home from work. I began to notice as we drove by some of her more affluent friend's homes, she would suddenly have to tie her shoe or pick up paper she had dropped on the floor. Could my little girl be **embarrassed** about riding in the car she found? Yes, she was. But this didn't last. Fortunately, it was just a phase. . .the first of many Rosie and I were to observe in our beautiful, maturing daughter.

The Isetta performed very well in its daily job of getting me to work. In the rare ice or snow storms in this part of the country, two cars were the only ones consistently on the road, the Isetta and the original Volkswagen beetles. But, maybe surprisingly, even in the late 1950's and early 60's, Isetta parts were not easy to come by. First there was a company called Fadex in New York. They were an authorized Isetta importer and had everything one could want. The only problem was having to order and wait a fairly long time for delivery. After many successful dealings with Fadex, one of my orders came back from a new address. Enclosed was the information that Fadex had gone under and Eisenhauer Motors in Pennsylvania had taken over as the Isetta parts distributor. Eisenhauer was a good supplier. They could supply anything needed and also offered a much better person-to-person relationship than Fadex. Eventually, though, as Isetta sales fell, Eisenhauer Motors began having problems. Correspondence with John Eisenhauer revealed that he was about to discontinue all support of the Isetta. He

"He told me it was the last Isetta imported into North America..."

told me he had one Isetta left which he needed to dispose of. He asked if I would be interested in buying it. He told me it was the last Isetta imported into North America by way of an east coast port. Since my 1957, bought in '58 Isetta was beginning to develop cable problems, rust in the gas tank, loose front end, battery box rotting out, and all the other little quirks we have grown to know and love, I made him an offer of $750.00 provided he would ship it from Pennsylvania to Alabama. He agreed and I became a two Isetta owner.

The delivery of the 1960 Isetta from Pennsylvania was an experience in itself. I knew it was coming by truck and sort of wondered how it would be unloaded, how I would be notified, etc. I found out when I got a call from the Kellam and Doster trucking terminal advising me that I had a little car in one of their trucks. They wanted me to come drive it out! I went to the terminal only a few blocks from my house to find my new Isetta, looking uncommonly lovely (I had forgotten how great a **new** Isetta looked) back about midway the trailer on the 18-wheeler. It was locked and the brake was on. . .the reason I had been called to come get it. Eisenhauer Motors had put it on the truck and sent me the key! I climbed in and felt like I was sitting on a stool. The springs in the new seat and the undisturbed "hay"stuffing in the seat had never been sat upon before. After driving my well

"I had forgotten the jerk of a new clutch . . . and the lovely smell of a fresh bug."

broken-in Isetta, I had forgotten the jerk of a new clutch, the tightness of the steering, the stiffness of the shift and the generally lovely smell of a fresh bug. This was, indeed, a **new** Isetta. The engine started immediately and I started easing out of the truck and onto the ramp to the unloading dock. Halfway down the ramp, it died. After running the battery down (it didn't take long. . .this **was** a new car) the obvious dawned, it was out of gas, even in the "reserve" position. Apparently, the gas had been dumped to make the truck trip safer. Rosie to the rescue went to a nearby filling station and came back with a jug of gasoline which got our new prize home.

Not too long after getting the new Isetta, I made one of my worst decisions. Since the original car was develolping a few problems, and since there was a local gentleman who wanted it, I sold it to him for $100. Realize, now, this was a running, everything working Isetta. The only thing about the transaction which made me feel a little bit better (and this is very uncharitable to say) was the engine blew up on him in less than a month.

From the early 1960's until the early 70's the second Isetta was used on a daily basis for local transportation. After this,

Continued on page 157

The Car Collector's Scrapbook

1958 BMW 600

by Gregory D. Wells

Photos by Ron Engel

The conclusion of World War II in Europe brought joy to almost everyone, but to the Bayerische Motoren Werke it marked the beginning of some very lean years. Initially utilizing its postwar personnel in the repair of vehicles belonging to the occupying forces, the BMW company eventually branched out into low-capital manufacturing, such as "pots and pans, building materials, airbrake compressors, baking equipment, agricultural machines, and bicycles," according to Ron Wakefield, co-author of the definitive book on BMW. In 1948, limited motorcycle production was resumed with the R 24 model.

BMW sought merger partners to assist in the production of their first postwar car, but no one approached was interested. So, on their own, the company developed the 501, a large, luxurious car meant for low production at a minimum of tooling expense. Models 502, 503, and 507 followed, the latter a lovely sports car with a 3.2-liter V-8. Some clue to BMW's postwar predicament can be gleaned from the fact that US importer Max Hoffmann's order for 5,000 BMW 507's had to be declined due to the lack of capital available for production. As Wakefield puts it, "All this folly — and from a business standpoint it was that — with luxury and sports cars was bringing BMW rapidly to the brink of new disaster. At the same time, the motorcycle business, which had gone so well earlier, was fading. Something had to be done."

First BMW took out a license to manufacture the Isetta, a bubble car designed by the Iso company headed by the Italian industrialist, Renzo Rivolta. They installed a 250 or (later) 300cc four-stroke of their own, replacing the Italian two-stroke, and clearly hoped that the little car, considered a size *below* sub-compact, would sell in the kind of numbers BMW desperately needed to stay afloat in the treacherous waters of the postwar automobile world.

The sad fact is that "Isetta was all right as far as it went; but it never got more than half-way to the bank," as assessed by automotive writer Michael Frostick. BMW's next move was to design the 600.

This new model, introduced in August, 1957, was an obvious Isetta derivation, retaining the forward opening door of that model. But the 600 was a four-wheel, four-passenger car and possessed another door on the right side for admittance to the rear seat. The 600 used an opposed two-cylinder engine very similar to the well-respected BMW motorcycle unit of the same configuration. This displaced 585cc and produced 19.5bhp, the latter representing a considerable detuning of the motorcycle engine. The reason for limiting the power output is to be found in the German vehicle taxation structure of the time.

Nevertheless, the performance of the 1215-pound 600 was thoroughly

excellent. Careful development and the twin's torquey nature caused all reviewers to marvel at the little car's smoothness and handling. *The Autocar* said that "the BMW can more than hold its own in an average main road traffic stream, while its compact dimensions and stability on the road enable an alert driver to thread it very quickly indeed through suburban or city traffic ...Cornering abilities of the BMW are exceptional." Even the American journal *Sports Cars Illustrated* opined "Of course acceleration is far from breathtaking but the gearing is so good and the engine so willing that there's a combination for every circumstance." With maximum speeds of 20mph in first, 36 in second, 52 in third, and approximately 65 in top, the BMW 600 was "a thoroughly practical, economical little vehicle built to the high engineering standards expected of its makers." Incidentally, economy was on the order of 50 miles per gallon!

Unfortunately for BMW, Germany had recovered from the war's devastation by the time the BMW 600 was introduced and the "upmarket Isetta" proved not to have been taken far enough. The front-opening door which carried the spare tire inside it for crush resistance and supported the swing-away steering column seemed to frighten buyers. Frantic BMW directors dropped the 600 in 1959 and introduced the Michelotti-styled 700. Its moderate success was instrumental in tiding BMW over until the introduction of the historic 1500.

In the 1500's planning stages, market research was conducted to assess the value of the BMW name and from this came the characteristics of the 1500 and later cars which led BMW into its glorious future. Incredibly, this research also revealed that the public did not consider the Isetta and 600 as BMWs at all, despite the obvious quality of the well-designed (if admittedly odd) little cars. But every BMW to this day owes a debt to the 600, for this car was the first to carry semi-trailing-arm independent rear suspension. In time, this "modern engineering coup" would find its way into other cars, including Mercedes-Benz and Rolls-Royce. Perhaps the BMW 600 was a market failure, but it would be a mistake to overlook its significance both to BMW and the entire automobile world.

* * * * *

The well-restored 1958 BMW 600 shown here is owned by Ron Engel of Davenport, Iowa, who bought the car two years ago and has completely restored it since. Total mileage on this coral red 'bubble car' is only 22,000 with one percent of that total having been driven since the restoration. Engel also owns an Isetta 300 that he is currently restoring.

ROCKY MOUNTAIN ISETTA CLUB MEET

A Super Success

by
Vic Cameron
Fort Collins, CO

On May 15th in Fort Collins, Colorado, an interesting phenomena took place: The Rocky Mountain Isetta Club hosted a meet where every single car was driven to the meet - a total of 5 Isettas! People were reportedly going nuts as they saw the Isettas driving down the highways, converging on Vic Cameron's home, site of this organization's first meet. All of the cars were brightly painted, new chrome, and ao,ost looked like Easter eggs rolling down the road.

By 1:30 p.m. we had all of the cars lined up on the front lawn, and everyone was meeting each other, comparing cars, notes, etc. Really having a good time! It made us feel very good that this was our first meet and 5 out of 8 members showed up with bubbles!

We took a total of 72 photos, and had plenty of spectators that stopped to look at these tiny cars. It was so thrilling; it reminded one of an artist displaying his life-long work or masterpiece. A great feeling! The meet lasted until 4:30 p.m. when one by one the members started their cars and drove home, proudly wearing their new club T-shirts of the Rocky Mountain Isetta Club.

The meet had been very organized with everyone showing up on time with clean cars. We plan to hold another meet in the near future, probably a pot luck picnic.

ATTENDEES OF MEET

1) Isetta 300 Deluxe, Red, **Art Cushing**
2) Isetta 300 Deluxe, Yellow, **Bill Anderson**
3) Isetta 300 Convertible, Green, **Sandy Jones**
4) BMW 600, Blue, **Harry Connors, Jr.**
5) Isetta 300, Green, **Vic Cameron**

Continued from page 155
because of the lack of spare parts and the press of getting our business into a new location, my Isetta fell on hard times for a few years. It had clutch problems, the steering was sloppy because of worn out rubber in the tie-rod ends, fuel valve leaking, the accelerator cable was - look, you know this list well. It's the same list **all** Isetta owner's have if they don't keep up with the problems as they develop. I not only lacked time for proper care of the bug, I lacked information about the existence of **anybody** else who also owned one of these cars! There just had to be some others out there, but where and how did I contact them?

When bi-centennial planning began in 1975, I got involved with the resurrection of the city's old 1914 American LaFrance fire engine. In order to locate parts for this beauty, we subscribed to Hemming's Motor News. In one of the issues, there was a small ad for the HMI Club in California which offered parts for Isettas. This confirmed there really were people left out there who owned and cared about their strange little cars. When the fire engine project was successfully finished after 6900 man-hours of intensive work, I was thoroughly convinced **anything** could be put back on the road if talent, time, and parts were available. Soon after the fire truck project was completed and I had become a member of the HMI Club, I had obtained the necessary parts to get my Isetta going and help from the fire truck crew to get it in pretty good shape. Since then, it has been used on a daily basis for transportation to and from work. I know there are "show car" people in our group and I say more power to them. I really wish I had two of these things so I could make one "pretty" and the other "utilitarian". For now I try to walk a reasonable line between a little car which will dependably get me to work and a collector's item which continues to appreciate in value as the years go by.

In case some of you Johnny-come-lately Isetta owners are wondering when the new will wear off, let me tell you. . .it won't. I'm as tickled as I can be that my grandchildren are having just as much fun riding in the Red Bug as their parents did many years ago. For over 25 years my Isettas have brought me pleasure, satisfaction, something for people to talk about, and definitely not the least, a mode of basic transportation which no other domestic or imported car has been able to offer. I just hope I don't see any of this current generation of kids nervously bending over to tie a shoe or pick up a piece of paper.

End

MILESTONES

A BIZARRE MUSEUM INTRODUCED ME TO THE ISETTA

(Everyone has their own story of how they first became acquainted with the Isetta, but the following account of this phenomenon is not your typical introduction to the wonderful world of bubblecars. For those of you who have seen the Isetta exhibit at the newly remodeled BMW FACTORY MUSEUM in Munich, Germany, the following article should bring a smile to your face. For those of you who haven't seen it, try to imagine being there by reading Lotar's impressions and experiences of what must be the most avant-guarde transporation museum in the world. -EDITOR)

Born in 1955, I had no real exposure to the Isetta cars of that era. Growing up in Southern California, the land of leisure therapy, I don't recall ever glimpsing any of these piquant pods in public. It wasn't until Oktoberfest 1980, while in Munich, Germany, quaffing massive quantities of pilsner, that I had the opportunity to appreciate the Dada design esteem of these bantam classics.

From the minute I set through the BMW museum door, I realized it wasn't going to be an ordinary queue of vintage vehicles. Every "respectable" mueum exhibits historical items; explains the function, importance and impact of these items on the environment. However, this exhibition was approached differently: rational information combined with emotional impression revealed a sense of candor and caprice (a BMW M1 painted by Warhol, a cabaret manikin with garters and spiked heels reclining in a provocative posture on the bonnet of a BMW 328, Nazi troops riding amidst the roar of bombs and crossfire on a BMW R75 motorcyle equipped with a machine gunned side car). In these display situations, "time" can be experienced. It is not so much the BMW product, but the idea behind the product that is the center of events; the theme being the concentrated history of the BMW concern.

This environment demonstrated that BMW has always been an integral part of society, directly involved with the motor sport events, with the cultural currents and the social and transportational developments of a unique epoch. The result is a sensible application of technology to obtain the highest degree of freedom and leisure fortitude required in our society.

by
LOTAR ZIESING
Santa Barbara, CA

The artistic concept was developed by the German Surrealist musician, Eberhard Schoerner, who stresses the maintenance of individual freedom. He was permitted this freedom to create the retrospective view over the last sixty years and the preview of the coming decade by means of a combination of the most deverse aids. He designed an environment that provides a direct, sensitive and tangible experience of the cultural, social and political trends of a motorized era.

After climbing through a circular barrage of sequenced events, the year 1956 hit me like a ton of smelt. Elvis, singing "Love Me Tender" from the cockpit of a bubble window contraption. What the hell is this, THING? How marvelous to idolize this creature of fantasy!

It was 1956. In the process of an economical boom, the motorcycle lost its importance as a means of transportation. Drivers wanted a roof over their heads and a wheel to turn; they wanted efficiency and little cars that resembled the playful, benign bubbles of their Teutonic consciousness. The Isetta was conceived and eugenically popularized. That was back in 1956.

Well, I said to myself, where can I find one now? This question plagued me during my travels through Europe, hoping to find one on the roof of an obscure Bratskeller or in the back seat of a Mercedes sedan, no such luck. My obsession grew after returning from Germany, until one day I noticed in our local paper: "1957 BMW Isetta, once owned by Ralph Nader, stored seven years, $650." Believe me, I was the first one there with the cash in hand at 10 p.m. that evening barging in on people that needed sleep.

Two years later, after a lengthy restoration, waiting for miscellaneous parts and other assorted mishaps, it's coming together. and boy, am I excited. Because I know what will be going through surprised minds when I drive down the main drag, pop that frontal lobe and exit, wearing nothing but Art.

THE BMW MUSEUM IN MUNICH, GERMANY

My Very Favorite Car

by
John Jensen
San Francisco, California

In 1957 I was out of high school a couple years, still living at home there in Council Bluffs, Iowa, and commuting across the Missouri to Omaha each day where I ran IBM machines for the Union Pacific. That job just barely provided me with enough income to sustain my "big Olds", a '54 98 with full power, white leather interior and dual pipes with a low rumble to cover a bit of valve clatter. As somewhat of a teenage connoisseur of automobile design, I looked upon that Olds as the state-of-the-art in Detroit flash and muscle. Here was the perfect conveyance for gliding along moonlit country gravel roads, or getting cozy at the drive-in on a warm summer night with your "steady".

By then I had become aware of a few imports, the MG for sure, plus the '56 Porsche that a friend owned. Of course, the appreciation of one of these little imports required a totally different point of view. My younger brother had a Triumph motorcycle, and I was learning to appreciate the single cylinder bikes prevalent at the time.

One day while shopping at Penney's in Omaha, I came upon a special display featuring a new white Isetta 300. The local Isetta dealer kept it there for several months. I stopped by that store on numerous occasions to examine that Isetta from every angle imaginable. I was totally fascinated by the unique design, and it made so much sense! Well, of course it wasn't quite right for my own lifestyle, but I was intrigued by it.

NEW ISETTA OWNER. PROUD, BUT HUMBLE - 1958

By '58 I decided to study engineering, with plans to move to Chicago. I drove my Olds over to the Volkswagon dealer in Omaha with a plan to trade for a new VW and get some cash back as well. Surprise, surprise! The dealer wanted my car and $300, plus a 3-4 month wait for delivery. Such an insult!

But a few days later, just out of the blue, the Isetta dealer in Omaha called to say he had a slightly used Isetta with a thousand miles on it that I should come at least take a look at, so I did. Sure was a clean little car, bright red, a sunroof, and the four speed stick "off the wall" was pretty exotic stuff for a guy that learned to drive on a '39 Chevy column shift. The dealer offered me the Isetta plus $500 cash for my Olds, and I took it!

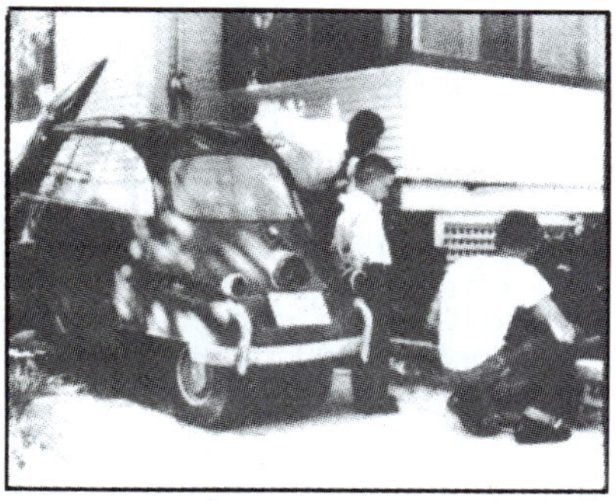

ENVIOUS FRIENDS OBSERVE MAINTENANCE TECHNIQUE.

Driving back to Council Bluffs that evening was a mysterious experience filled with lots of mixed emotions. Perhaps I had been a bit hasty, how were my family and friends going to feel about this thing, what was the real potential for ridicule here anyway? But, after a few miles, I was getting accustomed to shifting the reversed pattern with my left hand, enjoying the vibration of the engine, as well as the cornering responsé and solidity of the little jewel. Shucks, if this wasn't the next best thing to a Porsche, who cares!

And then, right there on Main Street at dusk, I stopped for a red light. This primo '57 Chevy stops along side, a shiny black one owned by the very coolest guy in town, and right next to him, the most desireable girl in town. I was maintaining my composure, trying to just act natural as the weird little ugly duckling Isetta sat there idling with a noticeable jiggle and a

PIN-STRIPES, KNEE-PAD & BURGER-TRAY. "300" LUXURY!

putt-putt-putt-putt. Seemed like the light was red for just hours , and I must surely be turning as red as my Isetta. Eager for the light to change, I slipped into first gear with a nice little clunk, and turned my head momentarily to my left to check-out the girl. I couldn't believe my eyes. Of course Mr. Cool could not permit a side glance of acceptance, but this adorable girl gave me a smile of total approval. The light changed, and they pulled away from me, but somehow cubic inches didn't seem to matter anymore. I had just discovered the power of "reverse status" and it felt great. As a status symbol, the Isetta was a viable alternative to a big V-8, and it looked to me like everything was going to be just fine.

My fears of being relegated to something less than respectability continued to dissolve. My family thought that a last I had done a sensible thing, my German girlfriend was impressed that my car was built in Bavaria, and my friends invited me to join their hot rod club, giving it sort of an international touch.

OF COURSE ISETTAS ARE COOL. YEAH, CRAZY MAN!

A few months later, on New Year's Day in 1959, I loaded all my essential possessions into the Isetta and headed to Chicago with four inches of fresh snow on the ground. It was a sunny 15 degrees F. with not much effect from the heater as I cruised across Iowa at 50 mph with a 400 lb. load. Took me almost two days to get to Chicago, but the Isetta performed just fine on this, the longest drive I had ever attempted.

During college in Chicago I found I didn't really need a car, and my Aunt Hermina permitted me to store it in the garage next to her "swept wing" Dodge. A few months before graduation I bought a new Sears battery, replaced a dried-out fuel tap, and resumed driving the Isetta in Chicago.

WHAT KIND OF MAN READS PLAYBOY - 1961.

One night in March there was surprise snowstorm. I said farewell to my girlfriend in the wee small hours and walked outside to find nearly a foot of snow on the ground at three in the morning. On Saturday night on the south side, I didn't want to have any car problems as I headed north through ghetto neighborhoods. No problem , the Isetta plowed right through the snow, and even on the Congress Ex pressway with the rear wheels pushing through six inches of mush, I made it all the way west to Oak Park and back to the frat house with no real problem. What a car!

Coming back across Iowa a few days later, I came upon a stretch of highway that had never been cleared of snow. The packed surface had turned to ice in the afternoon sun, and the Isetta felt a bit unstable at fifty. I slowed to 45, but a gust of wind caught me and I went into a wild spin, at least three full rotations before coming to rest safely at the right side of the road, adrenalin pumping as two farmers in a cattle truck passed me by and enjoyed a good laugh at the whole episode. It snowed and blowed all the way back home, drifting snow obscuring the road much of the time. But, with one's nose to the window aside the defroster vent, you can get an excellent view of the road, and the Isetta and I made it back to Council Bluffs safely.

"SEEKS ENGINEERING POSITION. WILL RELOCATE."

I enjoyed a couple months of job-hunting around Omaha before my friends and neighbors at the local draft board decided it was time for me to serve my country. I put the Isetta up on blocks in the driveway, covered it with a tarp, and had the pleasure of basic training in Ft. Ord at Monterey, California. In April, I stopped in Council Bluffs on my way to duty in Atlanta, Georgia, and prepped my Isetta for the 1,400 mile trip. By then

TUNING FOR OMAHA-TO-ATLANTA TRIP. 1962

I had a radio from a Mercury mounted on the floor just under my knees, speaker in back in a special enclosure beneath the parcel shelf, and polished aluminum plates over the vents in the front door to keep out unwanted drafts during the cold months. Nevertheless, a wool blanket was always there to put over the knees of driver and passenger. You kind of had to weave it around the steering column. I added a fold-down tray on the instrument panel. For hard cornering and bumps, I also had a small knee pad on the left side of the instrument panel to save me from repeated bruises.

Enroute to Atlanta I actually passed quite a number of vehicles on those two-lane highways. Outside of Knoxville, Tennessee, one night I was heading up one side of a mountain in third gear at 35 mph and passed a big truck. I didn't realize we were right at the summit as we started down the other side with this big rig gaining on me. There was just no way to pull out of his way as we raced down that winding two-lane black-top. I stepped hard on the gas and was soon pegged at 60 with him still right behind me and gaining. I decided to try neutral and coast faster if at all possible. It worked! The grade was steep enough that I eased on ahead of him doing maybe 70-75, cornering beautifully, and moved on out of danger. After that, I became real confident about the Isetta's ability to corner. What stability! Imagine what 500 cc's might do!

IN UNCLE SAM'S CARE. ATLANTA. SUMMER '62

The Isetta was perfect for me in Atlanta. With $68 a month salary, and a few off-post benefits, I could keep the gas tank filled and go cruising down Peachtree every night to strains of the Beach Boys and California Dreaming. That was the good life in '62.

That winter the Army decided to send me to Korea for a year. There was no time to drive the Isetta back to Iowa again, so somewhat reluctantly I sold my car to another soldier headed to D.C. I don't know where it might be now, but hopefully in the hands of an Isetta lover somewhere. It was a good one.

Since that time I've lived primarily in San Francisco, was an engineer, now a photographer. I forgot the Isetta for a long time, rarely ever saw one since '62, until in 1976 when my wife and I visited friends in Munich, and they took us to the BMW museum. There I saw a white Isetta, and it started me actually dreaming about Isettas, post Isetta syndrome or something like that. I had recurring dreams about an up-dated modern Isetta with double-wall stamped-steel construction, easily removable body panels, and always in a light rather chartreuse green.

In '83 I was coming out of the Coffee Tree restaurant in Vacaville, California, and there before me was an Isetta. I was ecstatic, taking my wife by the hand and pointing out all kinds of nuances to her. She was a bit less than interested. She dragged me back to our car, and as we headed to Sacramento, I started thinking . . . . . if that person has an Isetta, then there must be some others around, and surely, then, I can find one too.

It was several months before I lucked into Marilyn Felling and the HMI Club. But, with a membership roster I sent 40 letters to owners in Northern California, and finally bought a rust-free '58, almost identical to the red one I owned originally. I drove it home from San Jose on the freeway in the rain, operating the broken wiper with one hand, steering with the other, and trying not to notice the fumes coming through the various openings from the engine compartment. People passed me, mouths agape, smiling, pointing, laughing, giving me a 'thumbs up', right on, the macho types permitting only a peek out of the corner of their eye, trying to restrain their natural curiosity. Somehow, nothing has changed! Driving an Isetta is just like it always was, a wild and crazy experience! For me, it is just like 25 years ago, a trip back in time. Sometimes on the weekend, my wife and I will make a run for ice cream in the evening, and the folks ask the same quiestions and have the same amazement they had more than 20 years ago. We have a lot of fun with our car.

My rust-free California beauty is coming along nicely. I've managed to block all the fumes, fix the muffler, tighten-up the gear shift linkage, fix the wipers and the horn, sand-blast and repaint the wheels, and add new Michelin tires. I lucked into an extra engine, transmission and chain drive, just in case, and my next step is to rebuild the brakes before doing a big job on the body to make it like new.

This Isetta is fulfilling a lot of little dreams, and it still is my very favorite car!

MY FIRST ISETTA CHRISTMAS CARD.

"OF COURSE IT'S THE PERFECT CAR FOR A SKI WEEKEND, FROM BAVARIA YOU KNOW!" DEC. '83

SAN FRANCISCO, CA JULY '84
ISETTA #2 IS COMMING ALONG
JUST FINE.

The Isetta Sports

The Making of a Mini-"Q" Ship

by
Michael Dormer
San Diego, California

Miniature cars have always fascinated me. When I was a small boy I constantly badgered my parents to make the next family machine either an American Bantam or something along the lines of a Crosley "Covered-wagon" two-banger. "Economy, practicality, maneuverability, and besides, they're cute", I'd pitch them in my squeaky voice. Patient smiles, then irritation, then a sound whack on the side of the head was generally what I got for my incessant youthful filibusters. But they couldn't bang the automotive dream out of me. However, the message was clear. While growing up at home I'd have to abide by the rules and regs of the family and accept that ponderous Buick or Chevy as the logical "way to go" for the typical American motorist of the day. However, I knew in my soul that I was not to be the typical motorist of tomorrow, and the highways and byways of my imagination popped and buzzed with the sights and sounds of rocketing Fiat Topolinos and racy Morgan three-wheelers.

"I knew in my soul that I was not to be the typical motorist of tomorrow."

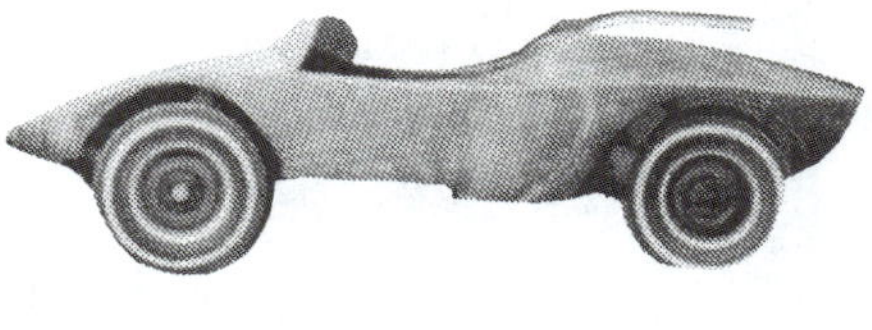

The first time I ever rode in an Isetta was in Boulder, Colorado, in the winter of 1958. Believe it or not, five of us crass kids jammed into the thing and chugged off through the snow and glare ice down to the local 3.2 beer joint. The noble, if somewhat overloaded, 300 held the treacherous road like glue, while domestic hogs spun off into the snowbanks and drainage ditches like so many greased bowling balls. I was sold. We were unstoppable, and I fell in love with the strange little invincible "egg". Then an idea came to me. Why not peel off the body from a wrecked Isetta and cloak the chassis in a sleek, low, two-seater fiberglass shell? Why, you'd probably have the world's lightest, smallest, most economical and just-plain-fun SPORTSCAR! The seed was planted, and I vowed that someday I'd do it.

Nearly a decade passed before I found her, hunkering in dishevelment in a weed patch next to a tacky apartment building. She'd been rolled and her once-prim flanks were as dimpled as a golf ball. The windshield appeared to have been used as a target by the local NRA chapter, and a recent heavy rainstorm had filled her to the gunwales with water. She was a pig, but all her parts were there. I located the owner, a rawboned sub-sailer, and we made the deal for $21. I pumped up the tires and pushed her home, and immediately stripped her down to the bare bones with a hatchet and a cold chisel, salvaging a few bits and pieces from the ruined body for later uses.

"Five of us crass kids jammed into the thing and chugged off through the snow..."

"... and began to sculpt the body using an array of strange tools like bread knives and cheese graters."

An industrial designer pal and I located a source of large, rigid polyfoam blocks, and purchased a small mountain of the stuff. He was undertaking a similar project, but chose instead a modified Volkswagen frame. We both decided to construct our bodies on the male mold principle, the only way to go for a "one-off".

I began by fastening the foam blocks to the chassis and began to sculpt the body using an array of strange tools like bread knives and cheese graters. Final smoothing was done with a Stanley Surform and various grades of sandpaper, then a few coats of fiberglass were laid on. Needless to say, I have many hours of sanding and adjusting ahead of me, and anybody who has worked with glass knows that it's damned awful itchy stuff to work with.

Because of the extreme lowness of the car, it was necessary to decrease the diameter of the steering wheel and widen the tube frame. I now sit between the rails, with maximum ground clearance of five inches. The decreased steering wheel diameter enables me to get in and out of the car, and quickens up the steering ratio a bit.

"She was a pig, but all her parts were there."

When finished, the car, in street-legal form, should weigh no more than 400 pounds wet or about half the weight of a factory Isetta. The weight reduction probably won't affect the top speed appreciably due to the stock stump-pulling final drive ratio, but acceleration times should be drastically improved. However, one must remember that my original concept was to create an ultra-compact, good looking mini-sportscar, and not a hyper-powered LSR-type machine.

Many more modifications and much more thought has to be put into my machine before she takes to the road, and frankly I'd be the first to discourage anyone else from tackling this type of project. The Isetta, as it stands, is a great and unique little machine, and when all is said and done, it's really sort of blasphemous to alter its character to the extent I have. In any case, I'll keep HMI members abreast of future developments concerning my project. Its completion is only dictated by the availability of my spare time, which lately seems non-existent.

"... and when all is said and done, it's really sort of blasphemous to alter its character to the extent that I have."

In Search of the Blues

There are easier ways to get to Beale Street than in a 1958 Isetta, but very few that offer so much opportunity for bad luck and trouble

BY PETER EGAN

IF THIS SAME cold autumn wind had been blowing through Chicago, they would have called it The Hawk—that famous lake-driven breeze that sweeps through the skyscraper canyons and sends a damp chill straight to the bone. We were not in Chicago, however.

We were in Rockford, Illinois, where, as far as I know, the wind has no name at all. Nevertheless, it was plenty cold and we were sitting in an aubergine-maroon 1958 BMW Isetta bubble car, watching leaves swirl around the street while we waited at a stoplight. The light, in fact, had just turned green, and there was a lot of honking behind us.

After a few moments, I looked over at my friend Chris Beebe, who was driving, and said, "Why don't we go?"

He looked back at me and said, "I think the throttle cable has slipped. I'm pushing on the gas pedal, but nothing's happening."

"Are you sure?"

"Of course I'm sure."

"You got the right pedal?"

"Yes. I have the right pedal."

I slid my side window open and listened to the air-compressorlike exhaust note of the single-cylinder BMW motorcycle engine behind us and said, "Maybe it just loaded up the sparkplug at the stoplight. I think it's responding now . . ."

Chris slowly put his head down on the steering wheel and started to laugh. Then I started to laugh too. Pretty soon we were speechless, wheezing with the kind of laughter you normally hear among people who have misplaced their Thorazine prescriptions. We were pretty tired.

"Only in an Isetta," Chris said at last, "could you have a serious discussion as to whether or not the throttle cable is connected."

The honking behind us persisted, so we opened up the front and only door of the car, stepped out into the cold street and pushed the car into a nearby driveway. We lifted one side up onto a curb so we could look under the car. Chris fiddled with the throttle cable, which had indeed slipped through its rubber retainer sleeve, and I looked at my watch with a flashlight.

"It's over," I said. "It's now officially too late to make it to Road Atlanta for

PHOTOS BY THE AUTHOR

the Runoffs. Even if we drive all night and nothing else goes wrong and we average 45 mph, we'll miss the last race on Sunday. Let's find a motel, get some sleep and figure out a new destination in the morning."

Chris nodded, caving in at last.

After five weeks of the usual relentless frame-up restoration—eating Doritos, losing sleep and waiting for the UPS truck to deliver rare parts from diverse corners of the universe—we'd lost our race with the clock by mere hours. For the first time in almost 20 years, we would not be going to the SCCA National Runoffs at Road Atlanta.

We found a nearby motel, checked in, unloaded our nominal luggage and collapsed on our beds. I can't remember if I took my shoes off or not. Meanwhile, outside our door, the Isetta sat between a Taurus wagon and a Chevy Caprice as if posing for one of those "What's wrong with this picture?" contests.

A fair question. What *was* wrong with this picture?

Why an Isetta?

Well, because Chris used to have one in college and thought it would be fun to have another one. He found this one, disassembled, in northern Illinois, bought it for $1000 and dragged it home. Chris and his wife, Dana, carried the Isetta into their basement, placing it next to the washer and dryer, to whom it appeared related. The car sat there for two years, until this fall, when I foolishly said, "We should rebuild that thing and drive it to the Runoffs at Road Atlanta next month."

So much for why. A more critical question in some readers' minds may be *what?* (Or, more accurately, "What the . . .?) As we would discover on this trip, some folks knew all about Isettas and had even owned them, while others had never heard the name and were virtually dumbfounded.

BMW Isettas were produced in Germany from 1955 to 1959, but the car was designed in the early Fifties not by BMW, but by Renzo Rivolta, a refrigerator manufacturer from Milan, which probably explains the front door and the heating system—more on this later. Diversifying from his line of Iso appliances, Rivolta decided to make scooters and small cars. The little egg-shape Isetta, originally powered by a 2-stroke engine, was essentially a kind of glorified Lambretta or Vespa, a scooter with weather protection.

Meanwhile, on the colder, darker side of the Alps, Germany was still digging its economy out of the wartime ruins, BMW was having a hard time selling its upscale 503s and 507s, and the Suez crisis had made fuel economy seem like a fine idea.

So as a stopgap measure to keep the assembly lines running, BMW bought production rights to the Isetta, reengineered it slightly and replaced the 2-stroke engine with its own R27 single-cylinder 4-stroke engine, bored out to a heady 295 cc, producing 13 bhp at 5200 rpm.

A 4-speed motorcycle transmission was used, with reverse added. The car had a short ladder frame with leading-link independent front suspension and quarter-elliptic leaf springs at the rear, with a solid rear axle/chain case bolted to the springs. The wheels were placed close together, so no differential was deemed necessary.

The body was of sheet steel, wrapped around a framework of steel tube, spot-welded together for a surprisingly rigid structure. The single front door swung outward, a cleverly articulated steering column swinging with it. There was no trunk, just a parcel shelf behind the 2-passenger seat. The whole package cruised along at about 45 mph and could be goaded into exceeding 50 at times.

Germans bought these cars because they needed them and because they were better than walking. Americans generally bought them because they thought they were hilariously funny, or because they were "cute" or "neat" or just interesting. Or, occasionally, even because they were better than walking.

Isettas suffered a lot of indignities on this side of the Atlantic. They were donated as booby prizes in contests, given away by car dealers as "a spare" and burned on top of homecoming bonfires. A body man who worked on Chris's car said that a girl in his high school class had one, and some kids set it on top of the school's air-conditioning unit. The car was destroyed when a wrecker operator used a sling through the open windows to lift it down, crushing the roof.

Chris said he used to drive his Isetta to work in Milwaukee, and the other mechanics would lift the back end as he tried to drive away, then drop it, which generally damaged the rubber donuts in the driveline.

What the hell? It was just a toy. Real cars were big, after all, impossible to lift—the more impossible, the better. Ah, we were a well-fed, healthy lot in those days, and the Isetta was the runt of the automotive litter.

I suppose to appreciate an Isetta, you have to spend five weeks restoring one, refurbishing and gazing upon each component. Only then do you realize the Isetta is not exactly a joke. No crude hardware-store project, it has dozens of specially built pieces and castings that are almost works of art. Look at the finely crafted steering box, the smoothly machined front hubs and brakes or the beautifully cast rear-axle housing, and you realize that all these were carefully engineered and produced by people who did absolutely the best work they knew how.

Stay up all night cleaning, painting and reassembling the elegant pedal cluster on an Isetta, and you would never want to see the car burned on top of a bonfire.

Nay, you would want to drive the thing all the way to Memphis.

And that is what Chris and I decided to do, the morning after our throttle problems in downtown Rockford.

We awoke well before noon—our first decent night's sleep in weeks—and ate a hearty breakfast without Doritos at a small cafe, swilling hot coffee and alternately looking at maps and our semi-trusty Isetta sitting in the cold parking lot.

"Where to?" Chris asked.

"South," I said, quoting U.S. Grant. I ran my finger over the pages of our road atlas and said, "Let's meander down the Mississippi River, deep into blues country. We could start over here at Hannibal, Missouri—Mark Twain's hometown—and head down to Memphis. Maybe check out Graceland and hear a good blues band on Beale Street. If Tom and Huck could make it downriver on a raft, we should be able to do it with a modern postwar conveyance such as an Isetta."

Chris looked doubtfully out the window and said, "Well, I hope it floats." Then he added, "One of my main goals on this trip—wherever we go—is to get through an entire tank of gas without having to repair the car. I could never do that with the Isetta I owned in college."

■■■■

The Isetta doesn't so much accelerate as accumulate speed, much the way a $20 U.S. savings bond accumulates interest.

■■■■

It seemed a small enough thing to ask, especially as the Isetta had only a 3.5-gallon gas tank. Yet we were less than 100 miles from home—a little more than two gallons into the trip—and we'd already made three roadside repairs. Most of these were teething problems, centered on the fuel system.

Chris had spurned the stock 22-mm Bing carburetor in favor of a great big honking 32-mm Mikuni motorcycle carburetor, "so it won't be as slow as the car I had in college." (Apparently, having an Isetta in youth leaves permanent scars.)

Unfortunately, he'd fashioned an intake manifold from a section of Peugeot radiator hose, which pulsed at idle like the lungs of a hummingbird and blew up after about 50 miles. We'd put in a new piece of hose on the road last night, this time wrapped with electrical tape. It was holding so far, but didn't bode well.

My first stint behind the wheel of the Isetta was quite enjoyable. Except for the left-handed, backward-H shift lever, which protrudes from the left wall of the cockpit, operation of the car is fairly conventional. The clutch works easily, the drum brakes (only one drum at the rear) are reasonably powerful, helped, of course, by the mere 800-lb. weight of the car.

Steering is medium-light, with good tracking and stability and no twitchiness or quirks, and the handling of the car is remarkably good. It zips through corners (when zipping speed can be at-

tained) with good grip and very little body roll, yet the ride is excellent for such a short car. There's a little hobbyhorsing over repetitive road bumps, but it soaks up potholes and railroad tracks that would shorten your spine in, say, a 1958 Sprite.

Power? Well, the Isetta doesn't so much accelerate as accumulate speed, much the way a $20 U.S. savings bond accumulates interest. It's almost fast enough not to be a hazard in city traffic, juddering up through the gears with a sound somewhat akin to a paint mixer. On the highway it hums along quite serenely at 45 mph, sounding more like a Taylorcraft with a 65-horse Continental. The whole car, in fact, has the feeling of a self-propelled airplane cockpit, and when you look out the windshield, you keep expecting to see a prop disc glistening in the sun. Rate of climb is poor, and uphill speeds can drop to 30 mph or less.

This is a good speed range, however, because nearly everyone can pass you effortlessly on a two-lane road. If you cruised just a little faster, say at 52 mph, traffic would back up. Only a few cars failed to pass us immediately—generally large sedans driven by aged farmers or freshly permed women in Church Lady glasses. Even these, fraught with indecision and uncertainty, would eventually creep by, chins forward, death grip on the wheel. In many cases, I am sure, we were the first car they had ever passed.

We sped down scenic Highway 2, along the Rock River, through Dixon, boyhood home of Ronald Reagan, then took 78 south toward the Illinois River, flowing toward the Mississippi through the fall color on a cold, crisp day. In Canton, we passed a supermarket with the Oscar Mayer Wienermobile in the parking lot, so we naturally turned in and parked next to it, hoping to set off a kind of psychedelic disorientation in the shopping public. As we parked, a woman poked her head in the window and asked, "Where *is* everything?"

"Like what, for instance?" Chris asked, grinning.

"Where's the front end? Where do your feet go? Where's the motor and all the things that a car has?"

A man in an MG sweatshirt walked up and said, "Man, I haven't seen an Isetta since I was in high school."

An older fellow looked the car over and said something we would hear repeated a thousand times on this trip, with no variation in wording: "I'd sure hate to hit a semi head-on in that thing."

As we drove south, Chris and I discussed which car, exactly, we would choose for the express purpose of hitting a semi head-on. We both agreed it should be something more substantial than an Isetta, perhaps a Nash Metropolitan or a Fiat 500. Crash safety was not a subject to dwell upon in the Isetta. Before the trip, we had even decided against installing seatbelts, theorizing that "it might be better to be found as far as possible from the scene of the accident."

Chris confessed that his worst safety nightmare was that we would hit the grille of a 1950 Studebaker, while I worried most about the front bumper of a 1953 Cadillac with the infamous spring-loaded "Dagmars." Either scenario was grim.

Highway 100 took us down the Illinois River into darkness, across Route 67 at Beardstown, which Chris pointed out was one better than Route 66. This comment led, after many hours of driving, to plans for a new TV program called *Route 65,* in which everything would be slightly disappointing,

■ ■ ■ ■

Only a few cars failed to pass us immediately—generally large sedans driven by aged farmers or freshly permed women in Church Lady glasses.

■ ■ ■ ■

compared with the original series:

Two guys, named Tud and Budd, would cross the country in an Isetta, rather than a Corvette, searching for adventure and Truth. Waitresses who had always wanted to "get out of this dump and see the world" would decline to ride with them, preferring to wait for a better car; small-town redneck sheriffs would refuse to arrest them for anything; they would be unable to pick a fight with guys at the factory, or even get a job there. Nelson Riddle's cousin, Billy, would write the theme music. Worst of all, carburetor problems would prevent them from ever leaving Illinois.

At which point our latest Peugeot radiator hose/intake manifold blew out, and we coasted into the parking lot of a motel in Pittsfield.

In the morning we found a much higher-quality intake hose (GM) at O.B. Dell's auto shop—with much kind help from O.B. himself—and headed toward Hannibal, stopping for lunch at Smith's Rockport Cafe in the little village of Rockport. Best lunch of the trip.

Crossing the Mississippi bridge, we dropped down into Hannibal and pulled up near Mark Twain's house, right next to the Becky Thatcher Book Store, where I bought hardcover replacements for a few of my disintegrating Twain paperbacks. More Isetta ballast. We then drove a few miles south of town, to Mark

Twain's Cave, fictitious hideout of Indian Joe, where Tom Sawyer and Becky Thatcher were lost.

Frankly, I had expected this to be a glorified hole in the wall, but it turned out to be an impressive and amazing labyrinth, miles of passages, chambers, connecting caverns and dead ends. You could easily get lost here forever, and some did. The skeletons of two local draft dodgers were found after the Civil War in a remote alcove. Many of Hannibal's children were temporarily lost in the cave, and organized search parties seem to have been a regular social function of the town.

Having grown up in a sleepy Midwestern town on a river, I identified closely with Tom Sawyer. I read the book at an impressionable age and spent one good summer of my childhood building rafts, fishing, smoking a pipe, swearing and sneaking out of the house at night. The one shortcoming of my hometown was that the only local cave was so small you could see the back of it from the highway, and so could your parents. Some hideout. So the vastness of the Mark Twain Cave was both a relief and a fulfillment.

Jesse James and his gang are also said to have hidden out in the Mark Twain Cave. But then, the James gang is reputed to have hidden out virtually everywhere in Missouri but the Lone Eagle Pub at the St. Louis airport, so a little skepticism may be in order.

From Hannibal we roller-coastered down Highway 79, the river road to Wentzville, where we found a motel. In the morning, we took small back roads south, stopping for a reverent moment at the front gates of Berry Park, Chuck Berry's rambling country estate near Wentzville. When you've spent as much time as I have playing "Carol" and "Little Queenie" on the guitar, you have to pay homage.

Meandering toward the Missouri River, we ran across Daniel Boone's last home in a remote valley near Defi-

ance. A lovely old stone and timber two-story house that Boone built by hand in his old age. Unfortunately, the simple, gracious homestead is now cluttered with a restaurant, a gift shop and the construction of a Living History Village. Poor Boone. Elbow room, indeed.

Some of the best sports-car roads in the world (if you had a sports car) led us into the wooded hills at the edge of the Ozarks, and we stopped for the night at Piedmont, after putting our muffler back on a few times and rejetting the carburetor with leaner jets from a Honda shop. In cruising through dozens of small towns, we noticed that reaction to the Isetta was nearly always the same, occurring in predictable stages: (1) accidental eye contact; (2) the classic theatrical double take; (3) "Isetta paralysis," in which kids stop dribbling basketballs, jaws drop, fuel fillers overflow and the thread of street conversation is lost in mid-sentence; (4) involuntary, reflexive pointing, where the arm comes up like a spring; (5) grinning; (6) calling to friends, frantically, to come and see the car before it's gone.

Everyone—and there were no exceptions on the entire trip—seemed to like the Isetta and thought it looked like a lot of fun, doubts about head-on collisions aside. A very common reaction was to ask a few questions, quietly examine the Isetta and then conclude, "This is all you really need to get around in."

South of Piedmont, our fifth day on the road, the weather warmed up for the first time. The sun came out and we were driving through bright fall color, side windows open. I'm told the Isetta can be a hot little greenhouse in summer sunlight, but that had not been a problem for us, so far.

The under-seat heat duct, which receives forced air from the engine fan via the shrouded cylinder fins, had proved frankly disappointing in the cold autumn weather. Apparently, most of the air molecules missed all contact with the warm engine and slipped through the fins untainted by thermal radiation. The result was something closer to refrigeration than heat, so we were able to travel without

Chris and I toured the Graceland mansion, which caused me to revive a line I used in a story long ago: "If bad taste were dynamite, you wouldn't want to smoke in here."

the usual Coleman cooler. Our sandwiches kept for days, and we ourselves didn't seem to age much, either.

Outside of Poplar Bluff, we hit a flat section of road and decided to do some timed acceleration runs. The results were impressive: 0–40 in 23 seconds; 0–45 in 32 sec.; 0–50 in 1 minute flat; 0–52 in 6 minutes. Fuel mileage had ranged in the mid-to-high 40s. Chris thought both acceleration and mileage would have been better with a stock carburetor.

Just north of Clarkton, we dropped down onto the flat, cotton-heavy plains of the Mississippi valley, and suddenly we were in the South. Delta country: warm sunny short-sleeve territory, with abandoned sharecropper shacks along the road, out of the old 40-acres-and-a-mule hill country and into the land of plantations, cropdusters and King Cotton. We'd crossed an invisible line of demarcation; Deep South, northern edge.

After a hard afternoon of driving, we cruised our little egg into Memphis, heading straight downtown and checking into the famous and ornate old Peabody Hotel, just a few blocks from Beale Street. We cleaned up and headed out for drinks, dinner and some music.

Beale Street is a famous old nightclub district that fell into disrepair years ago, but has been revitalized. I was here 14 years ago on a motorcycle trip, and the area was mostly bombed out and boarded up, except for a few bail-bond offices and liquor stores. Now it has dozens of blues clubs and restaurants, such as B.B. King's Club. B.B. himself got his start here in the late Forties, as "Blues Boy King," with his own radio show.

Chris and I ended up at the Rum Boogie Cafe, eating gumbo and listening to a superb guitar player named Preston Shannon sit in with the Rum Boogie Band. Shannon, at one point, asked people in the audience where they were from, and about two-thirds were from Europe, New Zealand, Japan and Australia. Memphis is a city of pilgrimage.

So in the morning we "pilgrimed" out to Graceland, Elvis' famous home. I had been here too, 14 years ago. By chance, I had arrived exactly one year after Elvis' supposed death, and the place was thronged with fans who lined up to parade past the graves of Elvis and his mother. The house was closed to tourism then, as family members were still living there. Now the place is a full-fledged tourist mecca, highly organized with tour guides and a huge museum across the street.

Chris and I toured the Graceland mansion, which caused me to revive a line I used in a story long ago: "If bad taste were dynamite, you wouldn't want to smoke in here." Not quite fair, of course, as none of us would care to have our tastes frozen back in the Seventies for all to see and ponder. And, as my brother, Brian, pointed out, "Elvis' strength was not really interior decoration, but he was a better rock 'n' roll artist than most."

Indeed. You read the names off the gold records on his wall, and you realize that the kid from Tupelo owns an awful lot of real estate in your brain.

Chris and I visited the Elvis car collection across the street, which, except for the staid conservatism of a few pink Cadillacs, was fairly glitzy. Elvis liked odd vehicles—3-wheelers, dune buggies and the like—with fiberglass bodywork and lots of metal flake. Given Elvis' taste for the unusual, we decided that he probably would have liked the Isetta. Had we parked it outside the Graceland gates when Elvis was still living at home, he might even have come outside to see it up close and maybe even have bought it from us, paying in cash. Then we could have flown home, avoiding Illinois.

As it was, we had to drive. And drive we did, straight as an arrow into the North, dialing summer back into a late, frozen fall with our small space capsule. Filling up as we left Memphis, Chris pointed out that we had actually used an entire tank of gas without fixing the car. A few miles down the road, our rubber intake manifold blew out and we had to adjust the points and retighten the muffler clamps, but that didn't matter. A record had been set.

We made it home in three days, with stops in Cairo and Kewanee, Illinois. A phantom engine miss plagued us all the way home, but the car kept running.

When Chris dropped me off at my house, we pulled into the driveway with the starter/generator brushes failing, the manifold collapsing and a bad sparkplug miss, but we made it.

Since then, Chris has taken the Isetta to his car shop and installed the old, original carburetor. The car goes faster now and gets better fuel mileage. He also found a short in the wiring and recrimped the muffler connection. If we made the trip again, it might be trouble-free. But neither of us has suggested it.

Still, I wouldn't hesitate to go anywhere in the Isetta again. It's quite comfortable, even for long hours behind the wheel; and long-distance travel at times has a best-seat-in-the-house charm about it, as though you are seeing America from an armchair in a bubble.

It's slow, but once you are resigned to that slowness, there's a humane aspect to the pace; no one surprises you, and you surprise no one on the road. Even the most dimwitted birds miss your windshield, farmers checking their mail along the highway regard you benignly, and their dogs and cats are safe. At 45 mph you can always stop, right now. Everything is avoidable; all accidents waiting to happen are still waiting. Probably forever.

Another part of the Isetta's charm, I suppose, is symbolic. One look at the car, with its severe economy of design, and you sense that it embodies a kind of lost freedom. No civilized nation would now permit an engineer to build a car this light, this simple, this eccentric or this logical for its specific purpose. It would be impossible to build and sell an Isetta today, without doubling its weight and bulking it out with bumpers, airbags, etc.

In any case, the Isetta makes a nice talisman from another era. It's like a Burgundy bottled before the spread of phylloxera in the 1870s; for better or for worse, we can't reproduce it and we'll probably never see its like again.

Not that anyone would want to build an Isetta again, unless resources and money should somehow become as scarce as they were in postwar Italy and Germany. We have 58-mpg economy cars now that cruise comfortably and quietly at 80 mph, carry four people and have heaters that work, so it would take catastrophically lean times, indeed, to bring us back to a design as Spartan as the Isetta's.

But then, the car was a product of catastrophically lean times. As Chris and I were reminded just before we began the Isetta's restoration.

Chris called an old friend named Klaus, in Germany, to see if he could find us a genuine Isetta shop manual. Chris and Klaus used to work together as mechanics, years ago, at a sports-car shop in California. When Klaus heard what we were doing, he said, "Oh no, Chris. Please don't tell me you have bought an Isetta."

"Why not?"

There was a long silence, and then Klaus said, "Americans think these cars are fun, but for Germans they are a reminder of very bad times, when we had nothing. We don't even like to see them."

He sent us a shop manual, nonetheless, and we were able to finish the car and take it for a drive down the Mississippi, into blues country.

For that purpose, the Isetta was something of a failure. We never did find a single person with the blues. We spent a solid week, driving all the way down to Memphis and back, and never saw anyone who wasn't smiling.

Still, I couldn't help thinking about Klaus's remarks during the trip. Would Americans have been so overjoyed with the Isetta if their own towns and garages were full of them? I wouldn't think so. In America as in Germany, it was the Isetta's rarity that created happiness and contentment. And vice versa. Cultivating the blues is a luxury, while living with them is something else entirely.

ISETTA JOHNS

FOR NEW AND USED ISETTA 300 PARTS

FAX 201 939 2208
P.O. BOX 451
RUTHERFORD, NJ. 07070, USA

871 BRY

BACK TO BACK

DOUBLE BUBBLE

The bubble car was a curious phenomenon of the fifties, and Peter Nunn has been out in two – an Isetta and a Heinkel

"Beep Beep! Beep Beep! His horn went Beep, Beep, Beep!"

Who sang that? Why the Headliners, of course, back in 1959, their *Bubble Car Song* telling the tale of a luckless limousine driver unable to shake off that funny little car following close behind. Cringing already? Too bad, there's more to come, as the *Bubble Car Song* ended, if memory serves correctly, with these immortal lines:

"Now I'm driving at 120, as fast as I can go. That Bubble Car was right beside, as if I was going slow. He pulled up by my window and shouted so I could hear – "Say buddy how do you get this car, out – of – second – *gear..?*"

That song, if nothing else, is a cogent reminder of the Bubble Car's irresistible – and unique – appeal that's proved such a boon for cartoonists, comedians and songwriters since the first 'bubble' was launched into an unsuspecting world in the early fifties. It could be argued that the 'bubble', having caught the public's imagination so decisively (think about it, *everyone* knows what a bubble car is) has now, like the Model T, become an accepted part of folklore. For a humble 55mph minicar with quirky looks, that's not bad going.

After the second world war, it took some while for industry throughout Europe to get back on its feet. In Britain, raw materials to make new cars were in short supply and fuel was also hard to track down. Cars, too, were thin on the ground which, inevitably, led to a black market and high prices for whatever transport was available. It thus became clear that a strong demand existed for a compact, easy-to-make car, that was not only sparing with fuel but readily available in the showrooms as well. It also had to be cheap to buy, tax and insure.

During the late forties and first few years of the fifties, a whole host of bizarre, motorbike-based 'specials' proliferated which fulfilled some of those needs (the French were conspicuously active in this respect), but it took until 1953 for the first *acceptable* minicar, or 'microcar' as the term is now, to be announced – and that was the Isetta, from Iso in Milan, the name standing for 'little Iso'.

Its concept was, in fact, familiar enough – a light, practical car with good economy, reasonable comfort and moderate performance. But what was so remarkable was the Isetta's shape; by doing away with a conventional bonnet and boot, Iso – a firm hitherto associated with scooters and motorcycles – had developed a car just 7ft 4½in long. That feature alone, although unusual, made it ideal for parking (the car was so short, it could be parked at right angles to the pavement!) yet the large, side-hinged door that formed virtually the entire front of the car was even more revolutionary. Not only that, it was the only door...

Air-cooled two-stroke power

The Isetta's bulbous body was mounted on a tubular chassis, power coming from an air-cooled 236cc, 9.5bhp two-stoke engine mounted transversely in the frame. The gearbox, in unit with the engine, passed the drive to the narrow-track rear wheels (19ins, in contrast to a front track of 45ins) and solid differential via an enclosed chain. Dubonnet bell crank suspension appeared at the front, with quarter elliptics at the rear, steering being by a worm and nut arrangement and braking by drums all round.

In 1954, BMW introducted their own Isetta which, though essentially the same as Iso's, sported a 247cc, 12bhp, four-stroke engine. Still a four-wheeler, this new Isetta sold well in Germany but, as local tax laws in other European countries favoured three-wheeled vehicles, Iso stopped manufacture the following year. The Isetta, now built under licence in Germany by BMW, in France by Velam and in Brazil (as a Romi Isetta), was offered in 298cc, 13bhp 'Export' guise by the Bavarians in '57, with sliding windows and swivelling quarter-lights. In 1957, Isetta assembly began in Britain, in an old railway building in Brighton, where the two versions – known in the UK as the '250' and '300' – were produced using a large number of British proprietary parts. But sales were slow due to the licencing laws which classed the Isetta among conventional four-wheel cars, so, two years later, Isetta GB Ltd introduced a three-wheel model to alleviate the situation. And it worked; sales, which took an upturn from that point on, increased still further with the adoption of right-hand-drive in the same year. The four-wheeler continued for export until the close of UK manufacture in '64 but the four-seat, two-door BMW 600 Isetta, built in Germany between 1957 and 1961, was never assembled in Brighton.

The Heinkel-Trojan story closely parallels that of the BMW Isetta, originating in Germany in 1954. Production at Ernst Heinkel's Stuttgart plant ended in '58, only to be taken up by a British licencee, Trojan Ltd of Croydon, in 1960, following a brief manufacturing spell in Northern Ireland. Assembly of Trojans, like the UK-built Isettas, tailed off during '64 although it is said that sales did continue into 1965.

Isettas started out as four-wheelers, progressing later to three-wheelers for licencing reasons, but with Heinkel it was the other way around. The first 'Cabin Cruiser' started out as a 174cc, 9.2bhp three-wheeler and it wasn't until later that 198cc and 204cc four-wheelers were announced by Trojan who, just to confuse matters, also offered a 198cc three-wheeler, known as the '200'.

Visually and technically, the Heinkel-Trojan and Isetta have much in common although to demonstrate the basic differences between the three and four-wheel 'bubbles', the specification chart compares a four-wheel Isetta with a three-wheel Heinkel. For the record, the three-wheel Isetta 300 had more power (!) than the lighter Cabin Cruiser, and was faster and cheaper as well. Yet fuel economy favoured the Heinkel which managed an average of no less than 73mpg (*Motorcycling* magazine recorded an astonishing 135mpg for a Heinkel travelling at a constant 30mph, back in 1961...).

Perhaps the principal difference between the Trojan and Isetta lies in the former's body construction. Yes, it's a quaint, bubble-shaped mini car with front-opening door but the Isetta has a chassis whereas the monocoque Trojan doesn't. While both cars share single cylinder ohv engines, four-speed

To the uninitiated, the Heinkel (left) and the Isetta (right) are barely indistinguishable. Their design and construction would give modern safety engineers a heart attack

'boxes and chain-driven rear wheels, the Trojan differs by having hydraulic front drum brakes with only a mechanical handbrake at the back. The trailing arm/coil spring rear suspension lay-out, along with the two turn lock-to-lock rack and pinion steering, is also different. Dubonnet ifs features on the Trojan but the 4.40 section tyres are skinnier.

The Heinkel is longer, narrower and lower than the Isetta but the extra length, and 1ft increase in wheelbase, goes to good use as it's a four-seater – of sorts. The front seats can take two adults with ease while the bench arrangement behind accommodates two children, or two grown-ups who don't mind craning forward for short journeys. The Isetta on the other hand, is a strict two-seater, with no room whatsoever for any back seat drivers. As compensation, there's a useful parcel shelf.

Contemporary tests showed the Isetta to be much the quicker of the two marques in terms of acceleration yet a top speed of around the 55mph mark seems to be a universal figure for all Trojans and Isettas – so quick they are not, but fuel thrifty they are with mpg figures of 60mpg or more being the norm rather than the exception.

Performance, of course, has always been one of the last reasons to buy an Isetta or Trojan. In the fifties, the bubble craze took off because of the cars' novelty value and comparative sophistication over the crude three-wheel minicars of the day. Suddenly the cheap and commercial 'bubbles' were fashionable wear for the well-heeled, and racing drivers also drove them with humour.

But with the arrival of such cars as the Fiat 500, Citroen 2CV and, latterly, the Mini, the practical advantages of buying a bubble began to be eroded. As the late Michael Sedgwick put it, 'the '57 Suez crisis and resultant fuel shortage granted a reprieve but it was growing affluence, coupled with the introduction of better, cheaper cars for the masses that finally dealt the death blow.' Yet fiscal regulations ensured that bubbles, many of them the worse for wear, made their presence felt on British roads until virtually the turn of the seventies. Today, enthusiasm for bubbles has been rekindled (perhaps it never went away in the first place?) thanks to the efforts of ferociously keen microcar club people in Britain and elsewhere.

In assembling this 'Back to Back,' we went right to the top. Mike and Helen Ayriss, the joint owners of the Isetta, have taken over the reins of the Register of Unusual Microcars (Helen is the secretary of RUM, a body aiming to track down all surviving micros) while Kelvin Luty, who professes to an incurable liking for bubbles of all sorts, runs the Trojan shown here, and is chairman of the Heinkel-Trojan Owners Club.

The Isetta and Trojan both date from 1962; they also share right-hand-drive and are three-wheelers. The Isetta is a '300', the Trojan a '200'. Kelvin recalls buying the Trojan for £10 (actually he bought two for £20...) while the Isetta replaced a Nobel that for four years served as Helen's everyday shopping car.

Representatives of the marques

To the man in the street, all bubbles are the same, yet side by side, the differences between these two are clearly visible. The Heinkel, with its greater glass area, looks more 'bubbly' than the Isetta which stands taller due to its separate chassis. Internally, the Isetta and Trojan are sparsely furnished with rudimentary controls, minimal instrumentation and hard bench seats; neither car is in tip-top condition but that doesn't matter one iota – both are representative of their marque and possess that inimitable bubble car character to good effect.

They say driving a bubble isn't easy but, with the Isetta, a novice need have little cause for worry. Once behind the ingenious steering column which pivots as you open the right-hinged door, the driving position resembles that of a conventional car. A normal H-pattern gearlever sprouts from the right and floor-mounted pedals sandwich the column. Switch on and the Isetta's 295cc 'single' throbs into life, sending vibrations shivering through the car. Once under way, the short-travel brake and clutch pedals work positively and the synchro'd gearchanges, deliberate and well-fashioned, are no problem. Steering, though, is disconcertingly vague despite light action and a useful lock; acceleration is modest yet the comfortable ride comes as an unexpected suprise. Top gear cruising is a noisy, bouncy affair and, overall, the Isetta sticks in the memory as a reasonably lively, unrefined fun car.

The Trojan, in comparison, has the feel of a slightly skittish go-kart. Inside the light, airy cabin, there's a quadrant-type gearchange close at hand ("that tends to put people off", says Kelvin). The driving position has the wheel close to your chest but legroom is fair. Visibility is excellent. The Trojan's engine, like that of the Isetta, vibrates noisily when revved yet, at tick-over, there's hardly a murmur. Starting, by turning the key and pumping the roller accelerator pedal, is a quaint feature but oh, that gearchange! An ill-adjustable cable on the day of our photo session gave poor Kelvin and yours truly all manner of problems – yet even when the 'box did function as it should, the changes were still a little tricky. The Trojan's stability, on the other hand, is good and the quick steering a delight. Ride is unusually well-damped, performance nippy for a 198cc car but brakes and clutch action are rather stiff.

So which is better? The Trojan is roomier, more economical and possibly greater fun to drive. You really feel you're in a bubble car while piloting or passengering in the '200' whereas the Isetta doesn't really convey the same sensation. (Whether that's a good thing or not, is open to debate!) Arguably, the Trojan is the better looking car of the pair, but its gearchange can create hassles and some microcar buffs maintain that its monococque construction is sometimes more of a hindrance than a blessing.

As for the Isetta, it's easier to drive and significantly quicker than the Trojan. It was also produced in far greater numbers (over 200,000 units in comparison to the combined Heinkel/Trojan total of well under 10,000) so finding one to buy today should be less of a headache. Aside from its woolly steering, the Isetta has few real vices as such, behaving far more like a conventional car than the Trojan. Typical prices for the two marques are about equal in the current 'classic car' climate and both cars enjoy the support of well-established car clubs, and carefully-orchestrated spares schemes.

Two micros, then, so similar in concept, style and appeal. Personally, I'd plump for the Trojan because of its looks and engaging bubble car style eccentricity but on a straight down-the-line comparison basis, there's little to choose between the two. Honours even it is then.

SPECIFICATION	Isetta 250	Heinkel Cabin Cruiser
Engine	Air-cooled, 'single'	Air cooled, 'single'
Bore x stroke	68mm x 68mm	60mm x 61.5mm
Capacity	247cc	174cc
Valves	Pushrod OHV	Pushrod OHV
Compression	6.8:1	7.4:1
Power	12bhp at 5800rpm	9.2bhp at 5700rpm
Transmission	Four-speed manual	Four-speed manual
Final drive	Solid axle, chain-drive	Solid axle, chain-drive
Brakes	Drums/drums	Hydraulic drums, front, mechanical, rear
Suspension F.	Dubonnet independent, coils, hydraulic dampers coils	Dubonnet independent, coils, hydraulic dampers
Suspension R.	Quarter elliptics, telescopic dampers	Dubonnet independent, coils, hydraulic dampers
Steering	Worm and nut	Rack and pinion
Body	Tubular box frame, separate body	Combined body and chassis
DIMENSIONS		
Length	7ft 6in	8ft 10in
Width	4ft 6¼in	4ft 5½in
Height	4ft 4½in	4ft 3½in
Wheelbase	4ft 10in	5ft 10in
Kerb weight	6¾cwt	5cwt
PERFORMANCE		
Max speed	56mph	52mph
0-40mph	16.2sec	29.2sec
Fuel con	62mpg	73mpg
PRODUCTION		
Years built	1953 - 1964	1955 - 1958
Nos built	200,000 plus	6400 plus
Price	£439 (1956)	£398 (1957)

More 'bubbly'? The Heinkel's curious curves

The Isetta's lines are less fussy and bulbous

The Heinkel has a fixed steering column

The Isetta's column swings out of the way